建筑立场系列丛书 No.20

C3

新医疗建筑
New Hospital(ity)

中文版

韩国C3出版公社 | 编

孙佳 陈思 王单单 王思锐 高文 王田竹 牛文佳 陈帅甫 | 译

大连理工大学出版社

建筑立场系列丛书 No. 20

C3

News

Archipulse

Digging Deep

New Hospital(ity)

Public Safety

STEVEN HOLL

海军码头

海军码头公司选择詹姆斯·科纳·菲尔德建筑事务所(JCFO)作为设计团队，来重新规划海军码头公共空间的未来发展。

这次比赛共有来自世界各地的52个团队参加，科纳及他的团队是入围海军码头区域设计团队的五支最终决赛队的其中一支。

在一百周年纪念的规划中，码头区域部分要求重新塑造码头的公共空间，包括码头西面入口处的入口公园、水晶花园、码头公园、东区花园、南部码头以及在码头周边区域分散着的较小的公共空间。

水晶花园

JCFO事务所设计了一个悬垂花园的壮观展示——从一个立面结构上悬挂了一系列大型植物盆栽。一端是大型瀑布的源头，而另一端是居住着色彩斑斓的奇异鸟的鸟巢。建筑师计划设10或12盆这种植物盆栽，每一个都覆盖着羊齿植物、苔藓、附生植物、藤本植物和其他植物，这些植物盆栽结构也可以被降低至地面高度，来形成一个戏剧化的立体花园。这种构思就是当地面要为特定的活动开放的时候，植物盆栽就可以被提升，在上方形成一个悬挂的花园；而在其他时段，一些盆栽可以被降低至地面上，作为不寻常的植物展览品，它们高高低低有可能形成内部洞穴；再者，所有的盆栽也可以落在地面上，成为绿岛、隐蔽处及微小花园的一种超现实的景观。

码头公园

码头公园与水晶花园相连，经重新翻修成为了一个使人更加愉悦的游玩场所。摩天轮和广受欢迎的旋转秋千也将会被重新修复。水晶花园和莎士比亚剧院两者的连接将得以加强并改善，可能会建一些高调的餐厅和咖啡厅，这样可以更好地协调并丰富周边地区的功能。

东区公园

在码头，最激动人心的地方无疑就是东区了，在这里大片令人惊叹的湖泊和天空沿着无尽的沙滩地平线相连，这与熙攘忙碌的城市形成了强烈的对比。这是一个非同寻常的地方，但是这里如今也会吸引相当数量的人们前往。JCFO事务所认为这里必须要有一个主要的景点，可以吸引更多的人来这里。为此，JCFO事务所提出建造四个独特的戏剧场所来帮助东区公园吸引更多的人群，这也理所当然地成了一个不得不看的目的地，有湖泊房间、浮动的水池、观景台和贝尔公园。

整个进展过程的下一个阶段就是由海军码头和JCFO事务所开始围绕这个项目诠释更多的含义。委员会强调说JCFO事务所的视角将会形成并指导码头区域的重新规划，但是很快指出最终的设计方案将会考虑到其他因素，包括其实践性、功能性、与海军码头公司的合作、委员会以及可用的资金。

Navy Pier _ James Corner Field Operations

Navy Pier, Inc. has chosen James Corner Field Operations(JCFO) as the design team with which it will work to reimagine the future of Navy Pier's public spaces.
Corner and his team were among the five finalists in Navy Pier's search for a Pier-scape design team with 52 submissions from around the world.
The Pierscape portion of the Centennial Vision plan calls for reimagining the Pier's public spaces including Gateway Park at the west entrance of the Pier, Crystal Garden, Pier Park, East End Park, the South Dock as well as the smaller public spaces that dot the length of the Pier.

Crystal Garden

JCFO propose a spectacular display of hanging gardens – a series of large scale vegetal pods that hang from an elevated structure. One might be the source of a dramatic waterfall, and another a bird-house with colorful exotic birds.

culture mile + the city lake mile

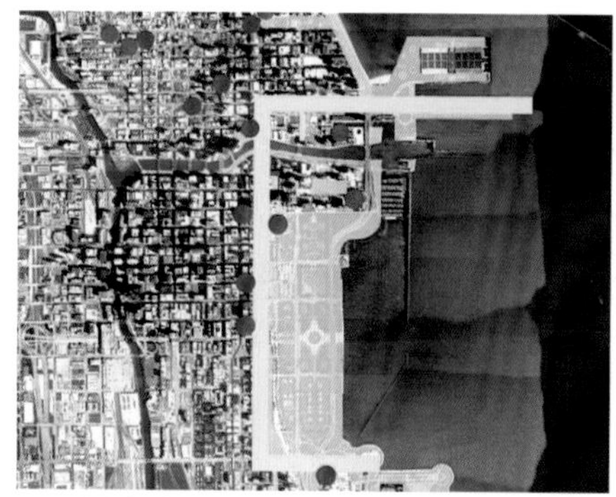
a new hub for Chicago

"plugging in" to the city

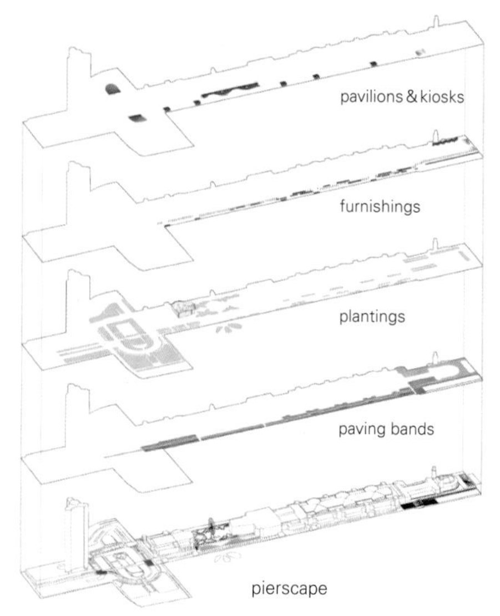

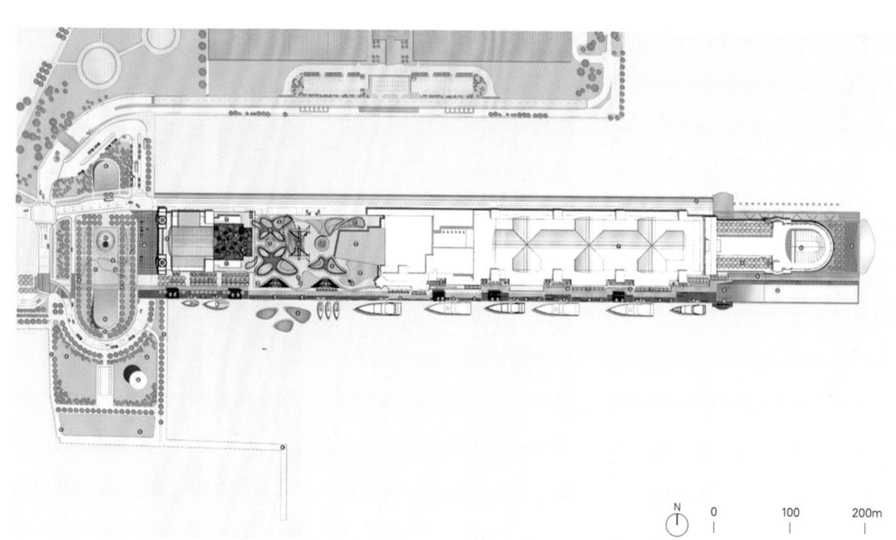

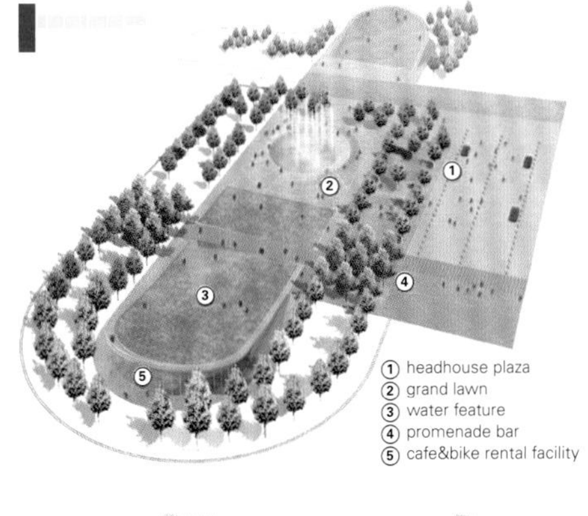

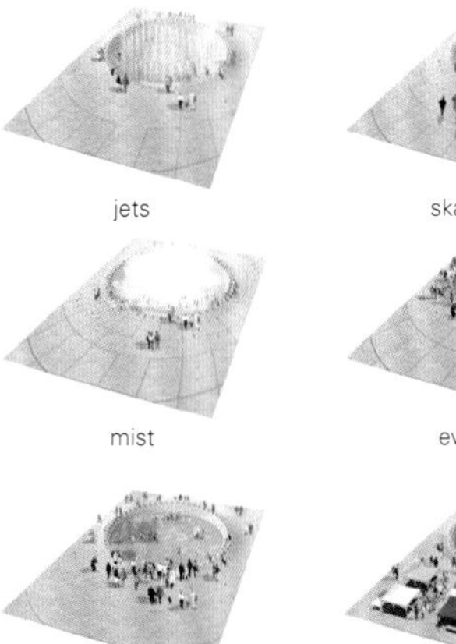

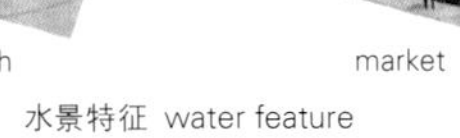

水景特征 water feature

动态感知 dynamic sensorium

metal roof panels

steel frame structure

polished stainless steel or rolled glass bonded to PVB panels

perforated polished stainless steel or rolled glass panes with steel tube structure behind

glass boat ticketing kiosk & canopy support

polished stainless steel

rolled glass

亭台结构&材料
pavilion structure & materials

Envisioning 10 or 12 of these vegetal pods, each covered with ferns, mosses, epiphytes, vines and other tectural plants, they may also be lowered to the ground level to create a dramatic spatial garden. The idea is that when the floor is required to be open for special events the pods may be raised creating a hanging garden above; while at other times some of the pods can be lowered and set on the ground as unusually and exquisitely planted exhibits with potential interior grottos; and at other times, all of the pods may be on the ground, as in a kind of surreal landscape of green islands, hideways and micro-gardens.

Pier Park

Adjacent and connected to the Crystal Gardens is the Pier Park, renovated as an even more enthralling space of play, fun, motion and buzz. The Ferris Wheel and the wildly popular swing ride will also be renovated. Connections and relationships to both the Crystal Gardens and to the Shakespeare Theater will be enhanced and improved, perhaps with active restaurant and cafe that better mediates and enriches uses around the perimeter.

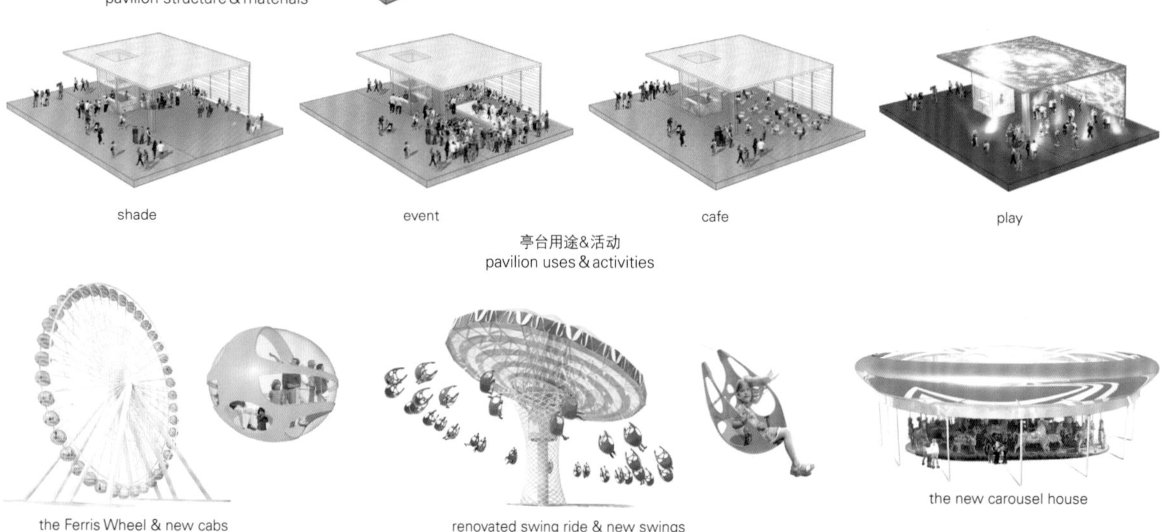

亭台用途&活动
pavilion uses & activities

码头公园景点
Pier Park attractions

项目名称：Navy Pier
地点：Chicago, USA
建筑师：James Corner Field Operations
主建筑师：James Corner
合作者主管：Lisa Switkin
项目经理：S arah Weidner Astheimer
高级设计师：Tsutomo Bessho
结构和环境工程师：Buro Happold
MEP和民用工程师：Primera Chicago
景观建筑师：James Corner Field Operations
照明设计师：L'Observatoire International
水景设计师和工程师：Fluidity
绿化墙和植物专家：Patrick Blanc
草坪专家：John Greenlee & Associates
工业设计：Billings Jackson
艺术知道和规划：Ed Marszewski
公共空间管理：ETM Associates
甲方：Navy Pier Inc.
竣工时间：2016 (expected)

East End Park

There is no doubt that perhaps the most dramatic space on the Pier is the East End, a place where the awesome scale of the lake and the sky meeting along an endless horizon sands in such spectacular contrast to the hustle and bustle of the city. It is an extraordinary space. But it is also a space that presently takes quite a haul for most people to get to. JCFO believe that there must be a major attractor, or set of attractors, that draw more people out to this remarkable place. To this end JCFO propose four unique and theatrical spaces to help ensure that the East End Park draws and attracts more people and becomes the must-see destination it already deserves to be: the Lake Room, The Floating Pool, Lakeview Steps and Beer Garden.

The next step in the process will be for Navy Pier and JCFO to begin to put more definition around the project. The Board emphasized that JCFO's vision would inform and guide the Pierscape redesign, but were quick to point out that the final design would reflect other factors including practicality, functionality, collaboration with the Navy Pier Inc. board, and available capital.

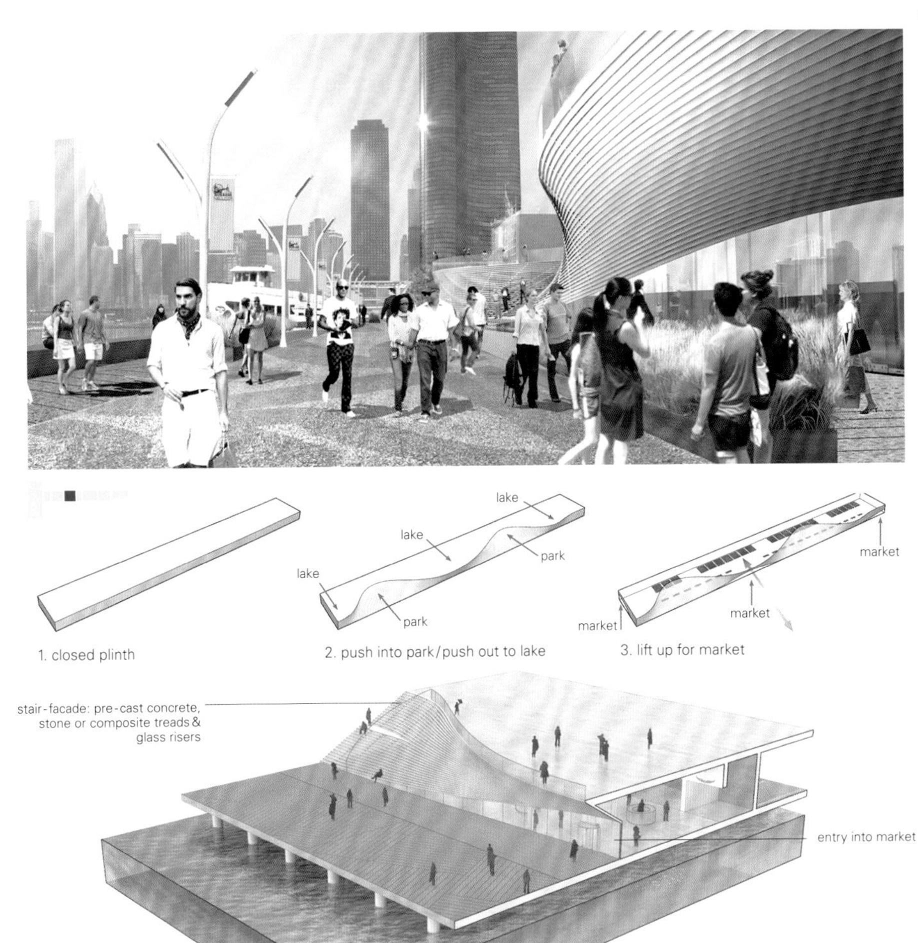

波尔多的文化中心——MÉCA

作为一个新文化中心的获胜项目提案，MÉCA围绕着一个朝向波尔多市和加伦河开放的公共空间筹划了现代艺术中心、行为艺术机构和文学与电影中心。该机构被构想成一个由公共空间和文化机构组成的单一环形结构，同时人行道抬升形成主厅屋顶，它沿着OARA的驿站塔垂直上升，将通道和FRAC自然采光的画廊连接起来，然后通过ECLA的档案室垂直返回地面，以此与海滨通道重新结合在一起。

该建筑的多种坡道和楼梯使其成为一种机构，面向公众开放，并且内外都让人感受到热情。具有城市生活特点的房间和楼梯上随意的座位使MÉCA成为一个充满活力的地方，也成为沿Quai de Paludate街和新式通道的生活的自然扩展部分。在城市节日或其他特殊场合里，MÉCA的外部可以转换成室外音乐会、戏剧表演或艺术装置的舞台。建筑本身和通道是由石灰石覆盖的，构成了波尔多建筑结构的主体部分。城市本身是用这种石头制成的，而所有通道和立面、楼梯和台阶，以及屋顶和天花板也都像是用同种材料雕刻出来的。

MÉCA, Cultural Center in Bordeaux _BIG

The winning proposal for a new cultural center, MÉCA arranges the contemporary art center, the performing arts institution and the center for literature and movies around a public space open towards the city of Bordeaux and the Garonne River. The building is conceived as a single loop of public space and cultural institutions as the pavement of the promenade rises to form the roof of the main lobbies, ascends vertically along the stage tower of OARA, bridges across the promenade with the sky lit galleries of the FRAC and returns vertically to the ground at the archives of the ECLA in order to reunite with the waterfront promenade.

The multiple ramps and stairs of the building create an institution that is publicly accessible and welcoming on the inside as well as the outside. The urban room and the informal seating of the stairs will make the MÉCA a lively place and a natural extension of the life along the Quai de Paludate street and the new promenade. During festivals or other special occasions in the city, the outside of the MÉCA can be transformed into a stage for outdoor concerts, theatrical spectacles or art installations.

The building and promenade is clad in limestone which constitutes the majority of Bordeaux's architecture. As if carved from the same material as the city itself – the stone is promenade and facade, stair and terrace, roof and ceiling all together.

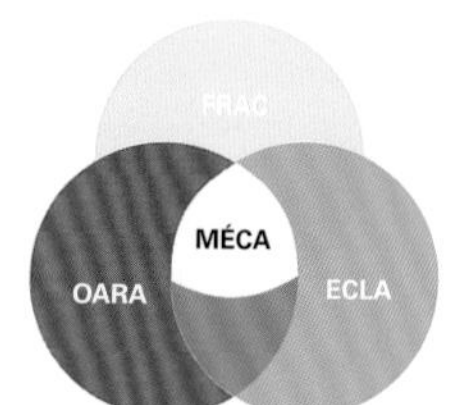

three institutions-one building

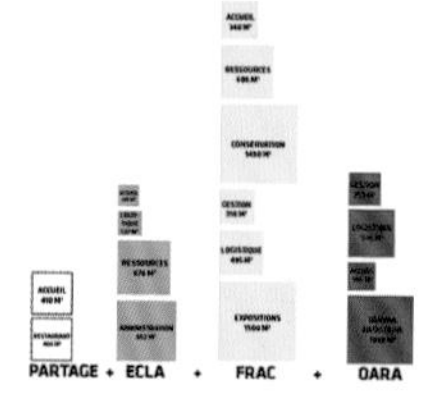

program

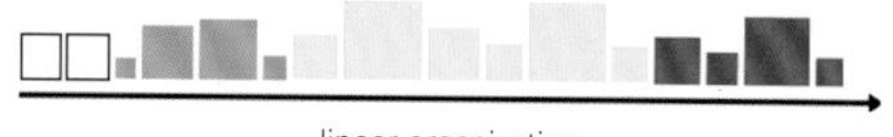
linear organization

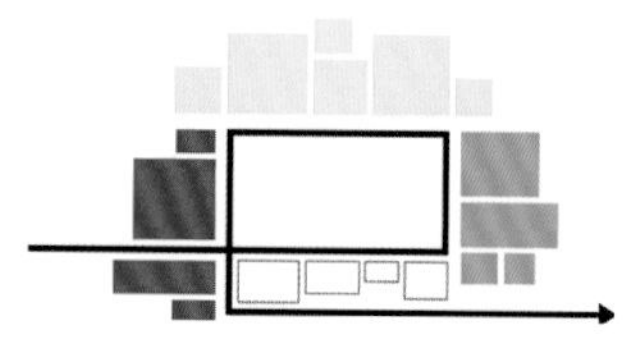
loop

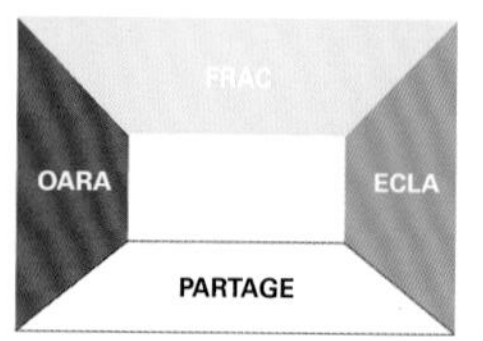

urban room

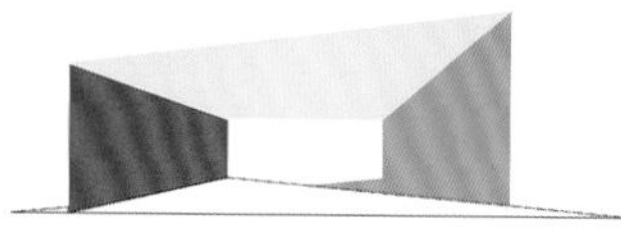
continuous promenade

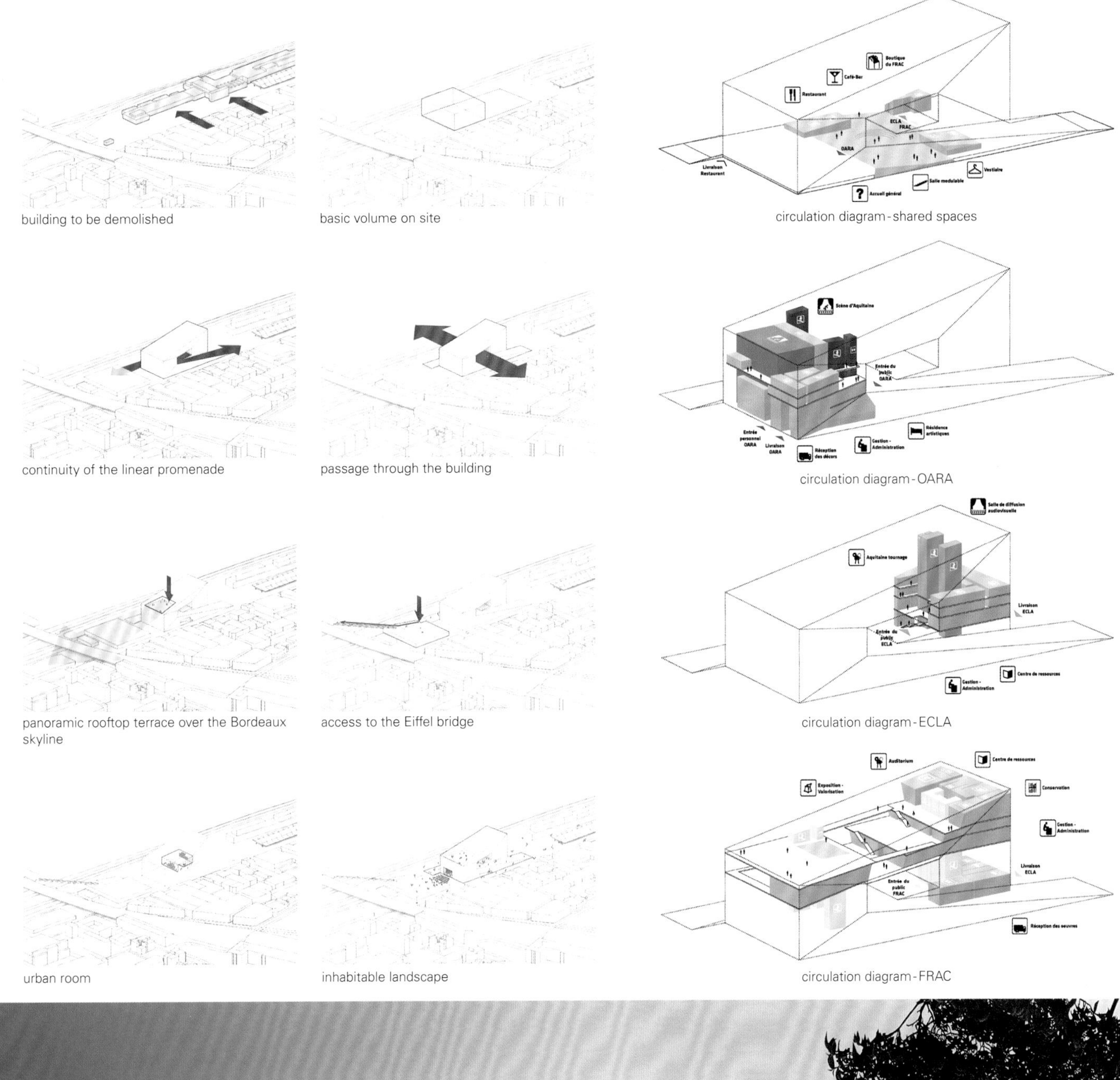

building to be demolished

basic volume on site

circulation diagram-shared spaces

continuity of the linear promenade

passage through the building

circulation diagram-OARA

panoramic rooftop terrace over the Bordeaux skyline

access to the Eiffel bridge

circulation diagram-ECLA

urban room

inhabitable landscape

circulation diagram-FRAC

oara

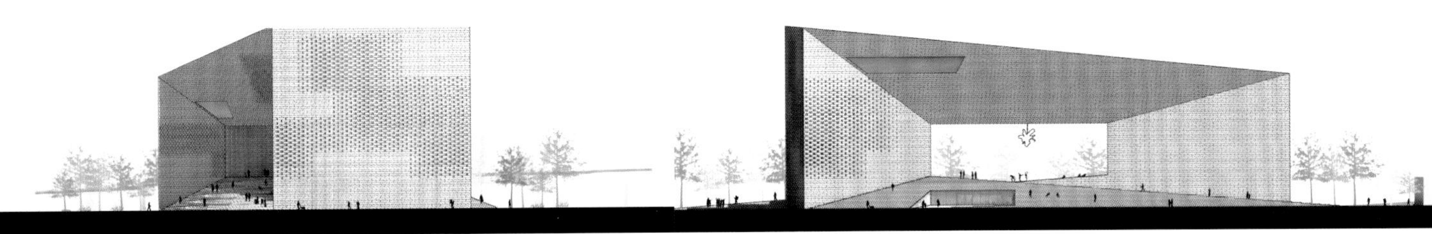

东南立面 south-east elevation

东北立面 north-east elevation

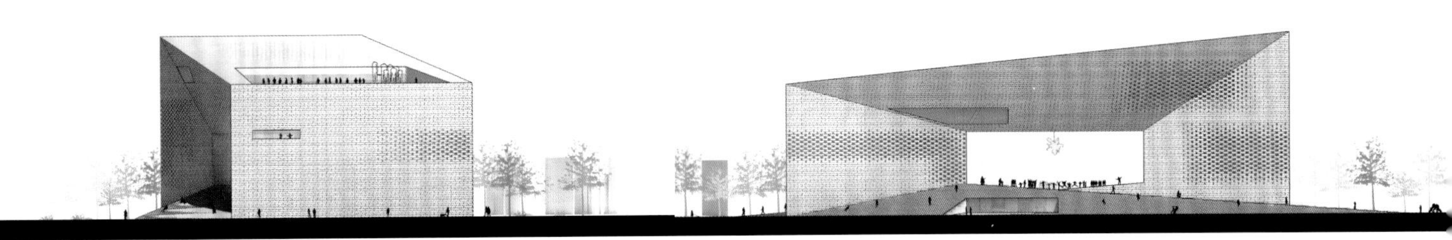

西北立面 north-west elevation

西南立面 south-west elevation

项目名称：MÉCA – Cutural Center in Bordeaux
地点：Bordeaux, France
合作者主管：Bjarke Ingels, Andreas Klok Pedersen
项目主管：Gabrielle Nadeau
项目建筑师：Jan Magasanik
项目团队：Édouard Champelle, Lorenzo Boddi, Yang Du, Karol Borkowski
合作者：FREAKS freearchitects, dUCKS scéno, Khephren Ingénierie, VPEAS, ALTO Ingénierie, Vincent Hedont, PBNL, Mryk & Moriceau, Ph.A
甲方：Conseil régional d'Aquitaine
用途：cultural center
建筑面积：12,350m²
设计时间：2012

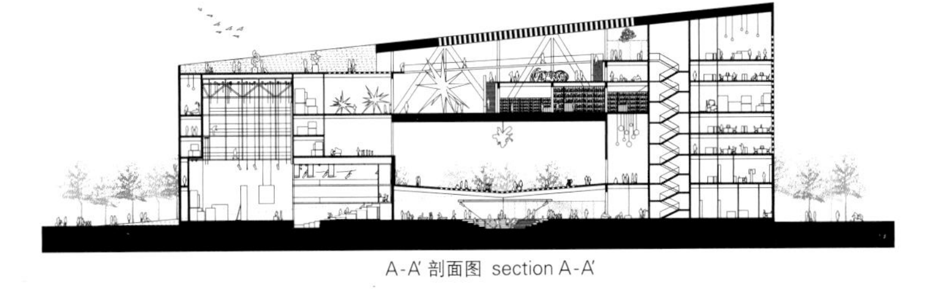

A-A' 剖面图 section A-A'

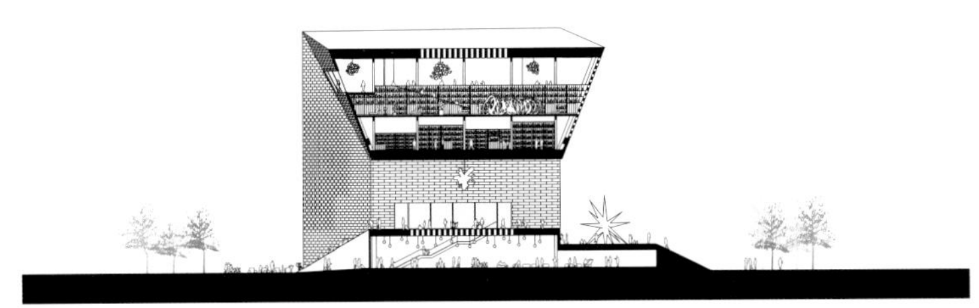

B-B' 剖面图 section B-B'

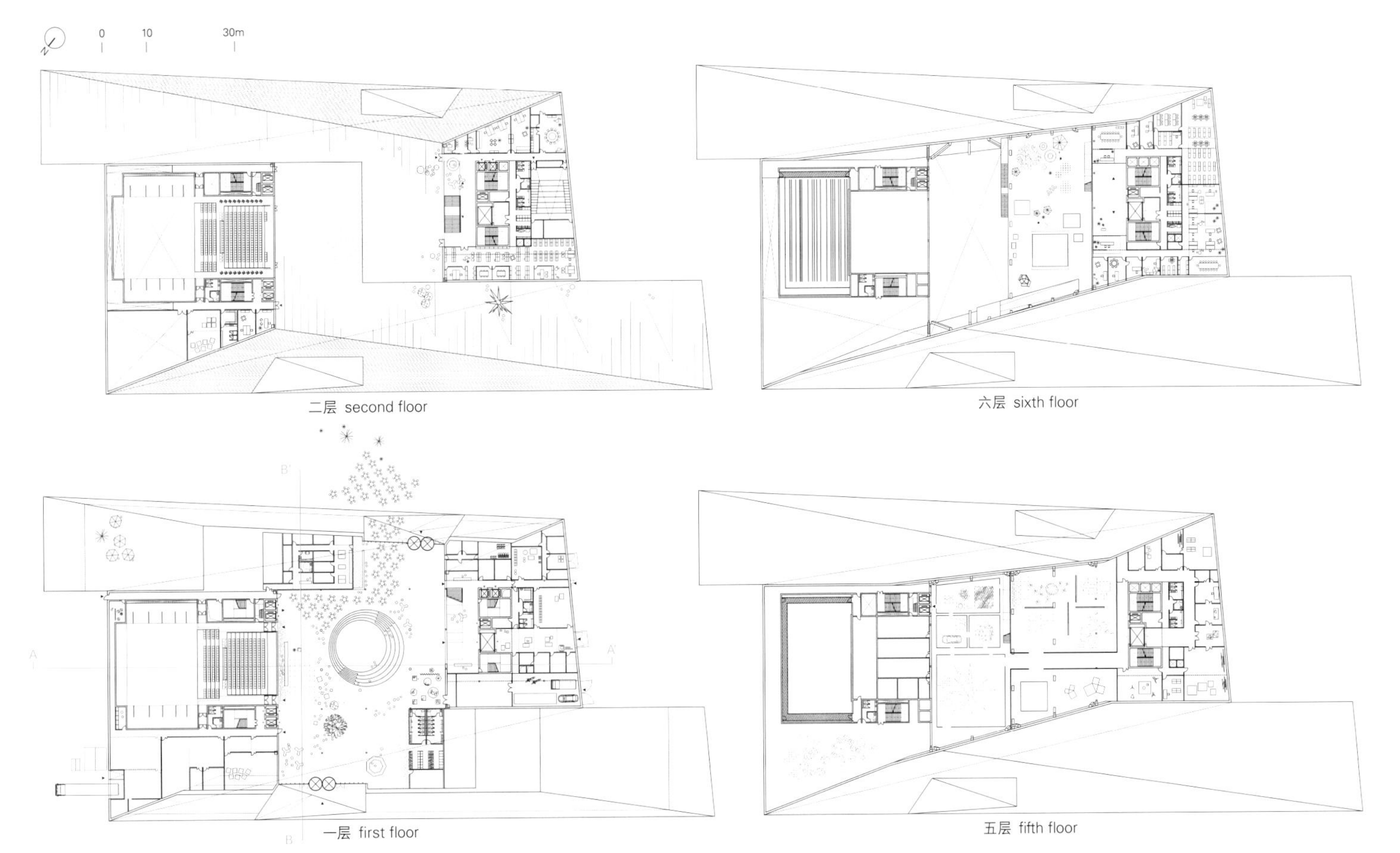

二层 second floor

六层 sixth floor

一层 first floor

五层 fifth floor

戛纳七屏展馆

2012戛纳电影节的开幕式上放映了坎耶·韦斯特的首部电影《严酷的夏季》，这个临时的金字塔展馆包含了一个由韦斯特创作团队Donda设计的七屏影院。

该展馆位于棕榈滩旁边，金字塔的顶盖边缘切开，展现出了戛纳和地中海的全景，同时在红地毯上面形成了一种漂浮的效果。连续的红地毯被不断加宽，最终延伸至200人座位的礼堂里，影迷们沿红地毯上升至金字塔里。OMA事务所与戛纳制作团队合作设计了一种最优化的三角形钢结构，可以悬挂屏幕、声音和放映等复杂装置。这样使得金字塔本身也隐隐约约地从戛纳白色的活动棚背景下显现出来。

该临时馆的设计使观众沉浸在一个由七个电影屏幕（大小约为5.2m×5.2m，5.2m×0.8m）所设定的空间里。《严酷的夏季》是在卡塔尔用七部定制的摄像机拍摄的，被视为放映的一个杰出代表。在这个空间里，通过七个屏幕的结构，观众被屏幕包围。这个项目是与Donda及2×4公司合作完成的。

7-Screen Pavilion Cannes _ OMA

Inaugurated at the 2012 Cannes Film Festival with a screening of Kanye West's debut short film *Cruel Summer*, this temporary pyramid pavilion contains a seven-screen cinema invented by West's creative team, Donda.

Located along Palm Beach, the pyramid's canopy is hemmed to open up a panoramic backdrop of Cannes and the

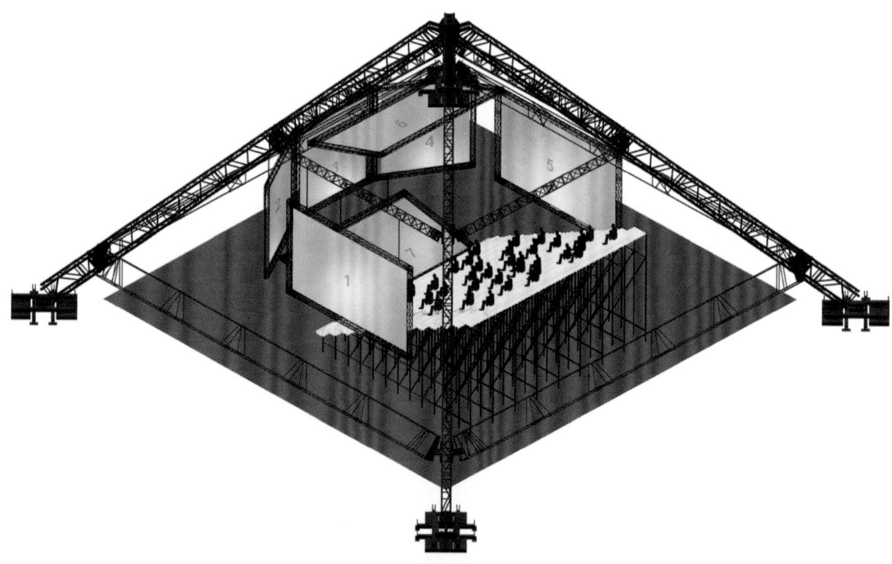

Mediterranean while creating an effect of levitation above the red carpet. Filmgoers ascend into the pyramid along a continuous red carpet that widens into the 200-seat auditorium. OMA worked with Cannes-based production team Ar'Scene to design an optimized triangular steel structure that suspends the complex screen, sound and projection installations. The resulting pyramid also subtly distinguishes itself from the context of white event sheds in Cannes.

The pavilion is designed to immerse the audience in a space defined by seven screens of cinematic proportions (17'x 17', 17'x30"). Shot with a custom seven-camera rig in Qatar, *Cruel Summer* was envisioned for this space as a constellation of projections that wrap around the audience through the configuration of the screens. The project was in collaboration with Donda and 2x4 Inc..

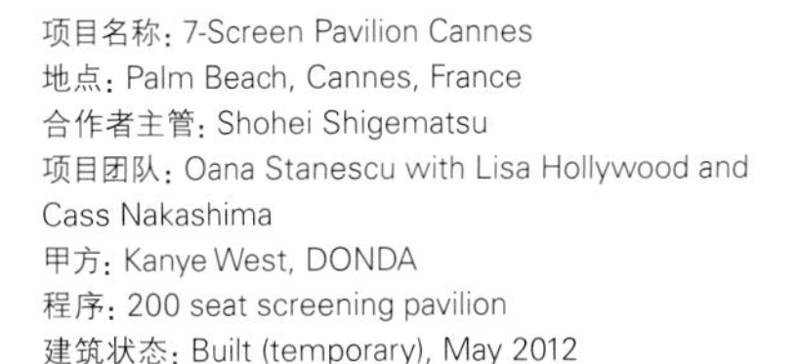

项目名称：7-Screen Pavilion Cannes
地点：Palm Beach, Cannes, France
合作者主管：Shohei Shigematsu
项目团队：Oana Stanescu with Lisa Hollywood and Cass Nakashima
甲方：Kanye West, DONDA
程序：200 seat screening pavilion
建筑状态：Built (temporary), May 2012
摄影师：©Philippe Ruault (courtesy of the architect)

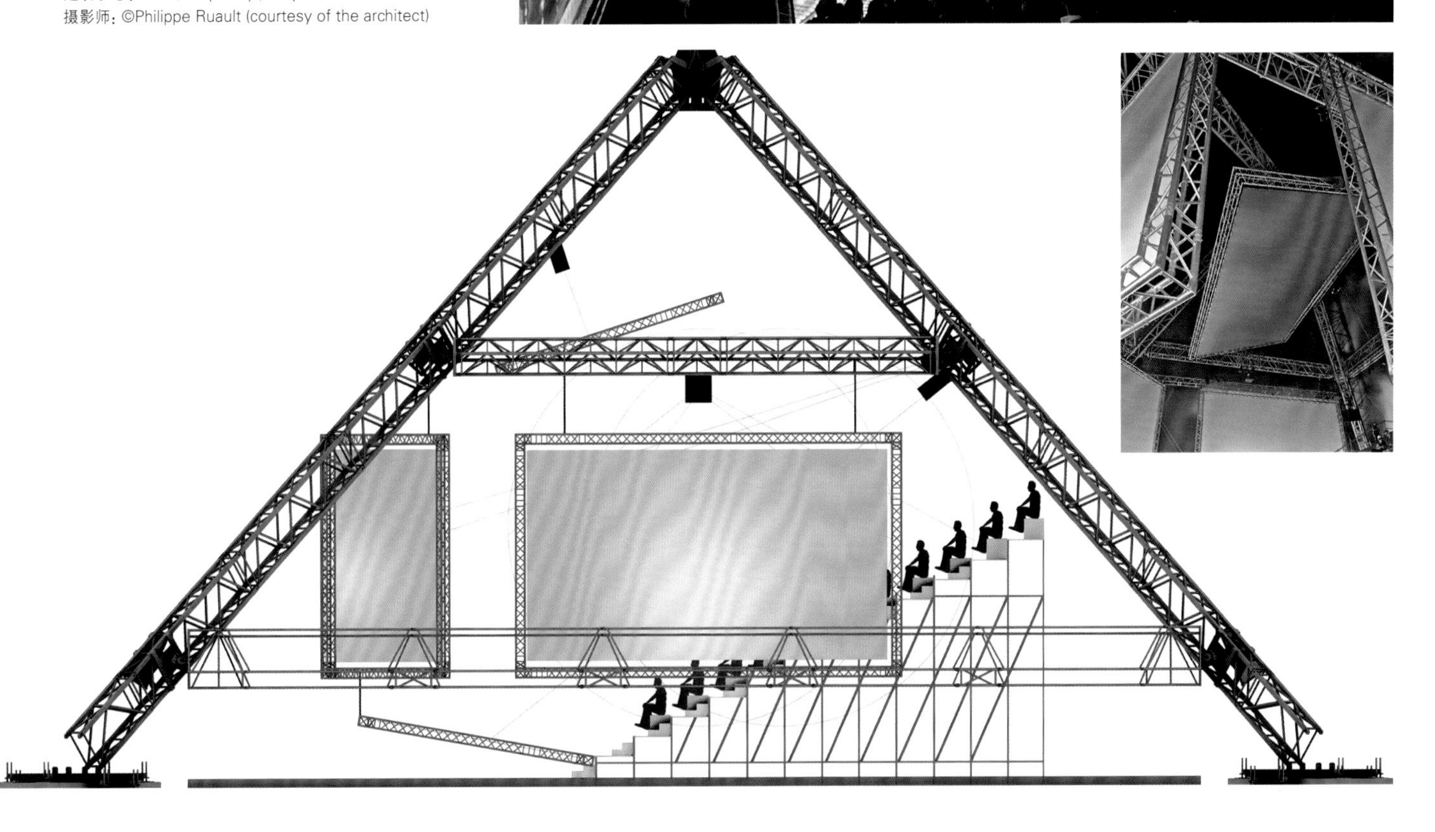

地下建筑
Digging Deep

2012蛇形画廊室外馆

Herzog & de Meuron + Ai Weiwei

从蛇形画廊自身所处的位置看过去，几乎看不到今年为它特别建造的展馆，除了那个悬浮在半空中的圆盘状水塘，该水塘构成了这座美术馆的屋顶。本着每年增建一座的原则，这已经是它的第12座展馆了。其屋顶比四周的草坪高出1.5m，这一高度对人们而言，无论是站着还是倚靠在其边缘都很舒服，如此美丽的一座雕塑令整个花园都变得灵动了起来。事实上，主体空间隐藏在下面的阴影里，这里沉入地下的深度也是1.5m。参观者向下进入到一片高低起伏的阶梯式景观中，显然，这一设计是通过把之前11座展馆的平面图叠加在一起而得到的。然后，设计师把某些部分凸显出来，构成了支撑纤薄屋顶的立柱、座位以及遍布整个空间但位于不同高度的小平台。越靠近中心位置，这些平台的高度越低。

经营者为这间展馆设有一系列的活动，其中包括现在众所周知的24小时马拉松赛事，就算不开展任何活动——小吃摊关闭，计划好的活动也不进行——这间展馆仍然是为数不多的、能够真正发挥作用的展馆之一。之前修建的一些展馆更像是小房子，而并非那种大而无用的建筑，因此到了晚上，当咖啡摊停止营业，但坐椅仍摆放在原处时，尽管美感仍在，却明显变得空荡荡，显得有些凄凉。

赫尔佐格和德·梅隆共同设计的这座展馆通过几种方式避免了"空咖啡馆"现象的出现。首先，其四面完全向外界开放，即使在个别地方，由于屋顶过低而导致进出展馆不方便时，它仍然与更加宽广的公园形成了明显的联系，视野也很清晰。同时，这种开放性也受到了人们的欢迎。每逢下雨，展馆里很快便会挤满了人，他们在下陷的空间里找地方安坐、倚靠或是聚在一处等待阵雨结束。对于那些想找地方就座的人而言，软木覆层同样很有吸引力——给人一种暖暖的感觉，即使偶尔经受了曝晒也绝不会变得太烫。这里是一个让人感觉轻松自在的地方。

这就涉及该展馆最引以为傲的方面了——设计者声称该展馆可以向人们揭示之前11座建筑的基础结构。这个设计理念本身似乎有些牵强——之前那些展馆根本就没有值得一提的基础结构，而且它们必然不会像赫尔佐格和德·梅隆设计的这座展馆挖得那样深。如此错综复杂的平面图叠加在一起构成了复杂的形状，即便是熟悉所有展馆平面图的人也无法从中解读出某一个展馆的占地范围。

然而，最终建成的空间效果非常好。与Snohetta于2007年设计的旋转屋顶相比，其内部空间没那么正式，更像是一个剧场，但如果举行活动的话，观众的注意力仍将汇聚到空间中央，而且随意摆放的坐椅便于让参观者拿来己用。与SANAA设计的极薄屋顶一样，这间展馆也成功地融入到了公园之中。恰恰是与公园融为一体的这个特点成就了那些最出色的展馆，巧合的是，也正是这一特点让它们成了富有鲜明特色的独立个体。

Serpentine Gallery Pavilion 2012

From the Serpentine Gallery itself, little can be seen of its twelfth annual pavilion, beyond the hovering, disc-shaped pool that forms its roof. Lifted a metre and a half above the surrounding lawn, it is at a comfortable height for people to stand and lean on the edge, a beautiful piece of sculpture to enliven this garden. In

Zaha Hadid, 2000

©Hélène Binet (courtesy of Serpentine Gallery)

Daniel Libeskind with Arup, 2001

©Hélène Binet (courtesy of Serpentine Gallery)

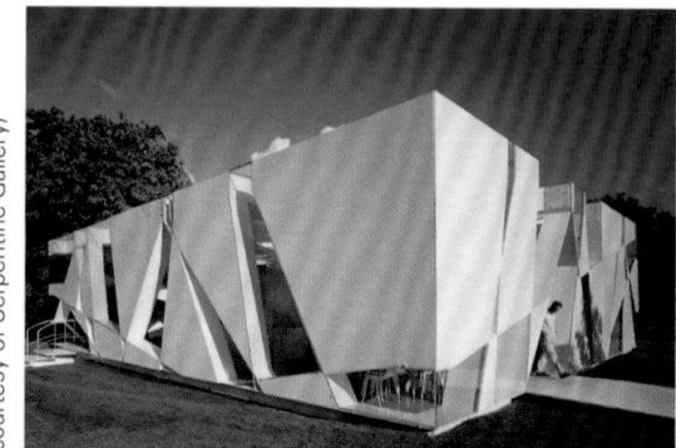

Toyo Ito and Cecil Balmond with Arup, 2002

©Sylvain Deleu (courtesy of Serpentine Gallery)

Oscar Niemeyer, 2003

©Sylvain Deleu (courtesy of Serpentine Gallery)

Álvaro Siza and Eduardo Souto de Moura with Cecil Balmond, Arup, 2005

©James Winspear (courtesy of Serpentine Gallery)

Rem Koolhaas and Cecil Balmond with Arup, 2006

©John Offenbach (courtesy of Serpentine Gallery)

Frank Gehry, 2008

©John Offenbach (courtesy of Serpentine Gallery)

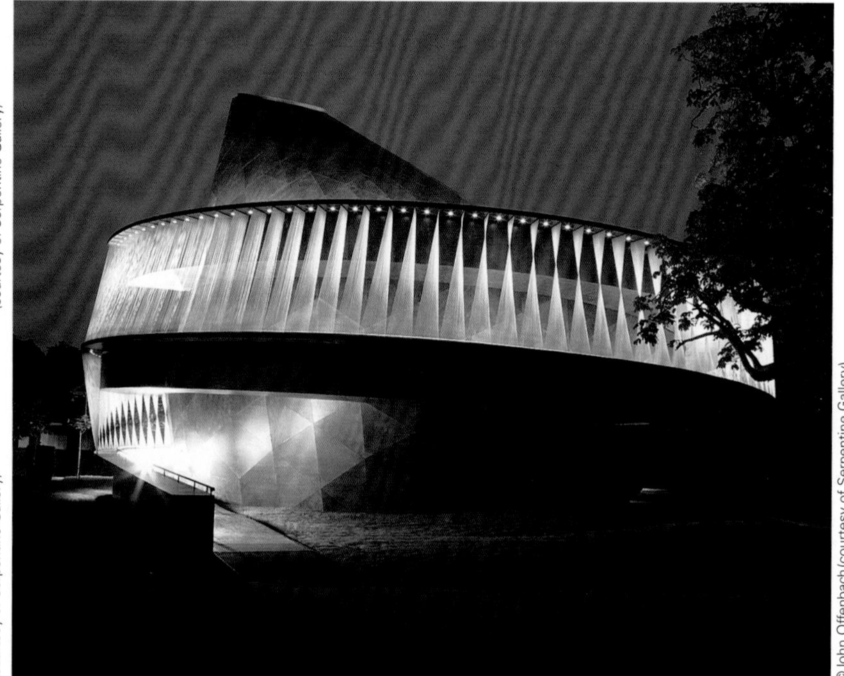

Olafur Eliasson and Kjetil Thorsen, 2007

©John Offenbach (courtesy of Serpentine Gallery)

Kazuyo Sejima and Ryue Nishizawa of SANAA, 2009

©Edmund Sumner (courtesy of Serpentine Gallery)

Jean Nouvel, 2010

©Philippe Ruault (courtesy of Serpentine Gallery)

Peter Zumthor, 2011

©John Offenbach (courtesy of Serpentine Gallery)

fact, the main space is hidden in the shadows below, submerged further a metre and a half into the site. Visitors descend into an unevenly stepped landscape, apparently generated through the superimposition of the plans of the preceding eleven pavilions. Parts of this pattern have then been extruded. This creates the columns which support the slim roof, the seating and the small platforms at different levels throughout the space, which get progressively lower towards the centre.

A series of activities is planned for the pavilion, including the now-familiar 24-hour marathon event, but even with little going on – the refreshment kiosk closed and no planned activities running – this is one of a handful of these pavilions that still truly works. Some earlier pavilions were more like small buildings than follies, so that with their seating still in place and the coffee kiosk shut up for the evening, they had the distinct air of empty cafes, their beauty as objects notwithstanding.

Herzog and de Meuron's pavilion avoids the empty-cafe scenario in a couple of ways. Firstly, its entire perimeter is open to the outside, creating a clear connection and line of sight to the wider park, even at points where the roof is too low to allow comfortable access to the interior. The openness is also welcoming. As it starts to rain the pavilion quickly fills with people, finding places to sit, lean or gather in the sunken interior as they wait for the shower to pass. The cork cladding is also compelling for those looking for somewhere to sit – warm, without ever getting too hot in the occasional bursts of sunshine. It feels like a very easy place to be.

This brings us to the main conceit of the pavilion – that it claims to be uncovering the foundations of the previous eleven years' constructions. As a concept it seems a little contrived – the pavilions didn't have foundations to speak of, and they certainly wouldn't have gone as deep as Herzog and de Meuron's sunken space suggests. It also isn't possible to read the individual pavilions' footprints in the complex shapes created by the superimposition of so many intricate patterns, even for someone familiar with them all.

The resulting space however, works really well. It is less formal and theatre-like than the interior space of Snohetta's spinning-top of 2007, but attention can still be focused towards the centre as events demand and the informally arranged seating is easily appropriated by visitors. It reaches out to the park as successfully as SANAA's wafer thin roof did. It is this connection to the park that has marked some of the best pavilions, this link ironically enabling them to stand alone as objects in their own right. Alison Killing

以前展馆的形状
previous pavilion footprints

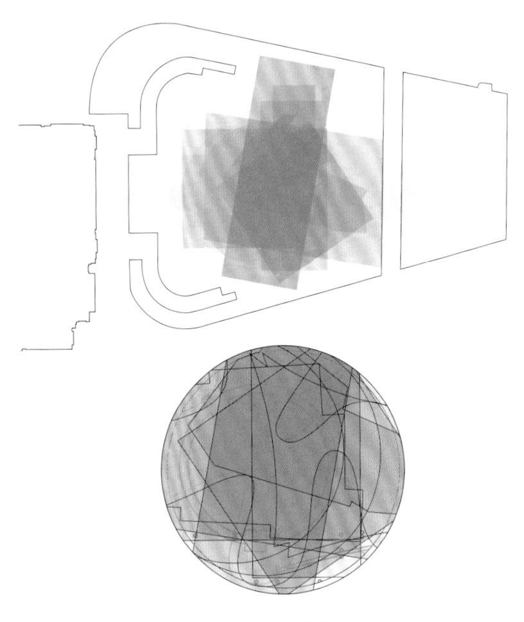

将各展馆的形状相叠加
overlay of pavilion footprints

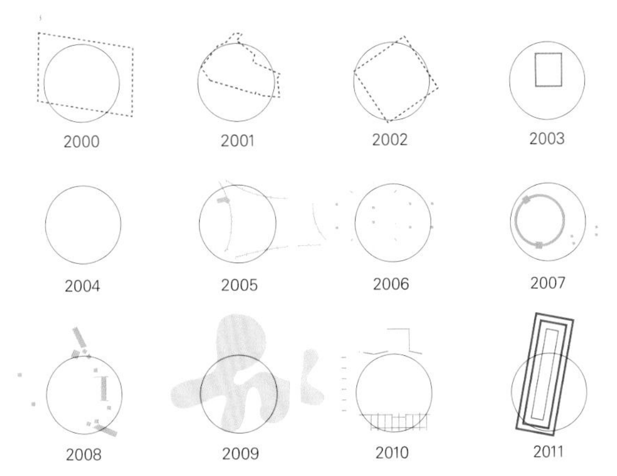

以前展馆的基础
previous pavilion foundations

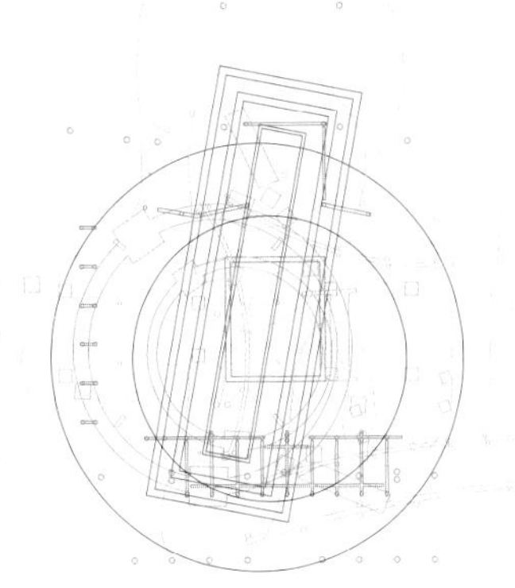

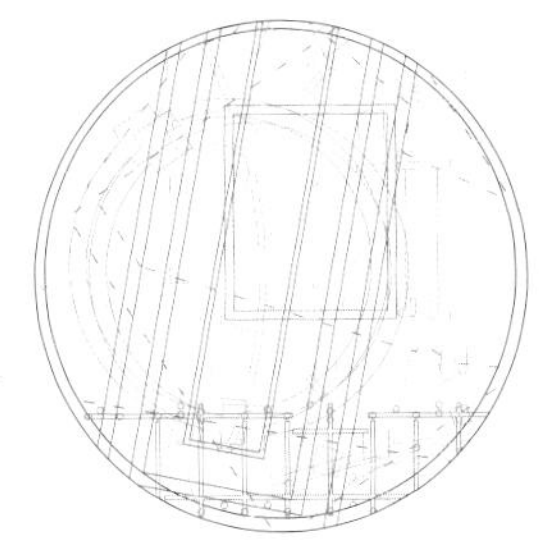

将各基础相叠加
overlay of foundations

自2000年起，每年都由一位不同的建筑师负责营建肯辛顿花园内蛇形画廊的夏季室外展馆。迄今为止已经有十一座，而建筑师在建的是第十二座。如此多的室外馆已经被设计建造完成，它们的形状是如此多样，采用的材料是如此不同，有鉴于此，建筑师本能地试图回避开这个无法回避的问题：他们最终会建造出一个实体——一座有形的展馆。建筑师另辟蹊径，在公园里向下挖掘了约1.5m，直到触及地下水。在那里他们挖了一个水孔，类似水井的结构，用来汇集所有落在该馆范围之内的伦敦雨露。如此一来，他们便将公园环境中的一个无形方面——地下水——融合到他们的展馆之中来。在建筑师向下挖掘到地下水的过程中，遇到了各种各样的人造物品，如电话线、残余的旧地基和回填材料。他们像考古队一样鉴别这些残余物质，确定它们是2000年至2011年间修建的十一座室外展馆的遗迹。那些展馆残余物的形状各异：有圆的，有狭长形的，有波点状的，还有被填平了的巨大建筑空洞。这些残余证明了之前几座展馆的存在，也证明它们都或多或少地对公园的自然环境造成了侵扰。

现在，之前展馆留下的所有痕迹都将显现出来，接受重塑。之前的地基和足迹形成了一团纠结的回旋线，好似一种缝纫花样。一片特色鲜明的景观就此形成了，这绝非任何人工设计所能匹敌的；其形式形状实为妙手偶得。这片景观的灵活性令人叹为观止，是安坐、站立、躺卧的绝佳位置，就算只是欣赏也让人心生敬畏。换言之，理想环境就是使游人们在蛇形画廊室外馆中可以继续过去十一年的所作所为。展馆内部采用软木镶面——一种触感、嗅感极佳的天然材料，适用于雕刻、切割、塑形等多种处理方式。

在每一座展馆的地基之上，建筑师都树立起一个新的结构（支撑结构、墙壁、片状结构）作为荷载组件以支撑展馆屋顶——总共有十一个支撑结构，此外，建筑师自己增加了一根可以随意摆放的立柱，就好像添了一张万能牌。屋顶效仿考古遗址上的屋顶。它悬浮在公园草坪上方几英尺的位置上，以便所有游人都能看到上面的水，水面映照出伦敦变幻莫测、广袤无垠的天空。如有特别活动安排，可以像放空浴缸一样把屋顶上的水放掉，让它重新流回到水孔中——整个展馆的最低点。之后，抽干的屋顶就可以作舞池使用，或者只是简单地作为悬置在公园上方的一处平台使用。

Every year since 2000, a different architect has been responsible for creating the Serpentine Gallery's Summer Pavilion for Kensington Gardens. That makes eleven pavilions so far, and our contribution is the twelfth. So many pavilions in so many different shapes and out of so many different materials have been conceived and built that we tried instinctively to sidestep the unavoidable problem of creating an object, a concrete shape. Our path to an alternative solution involves digging down some five feet into the soil of the park until we reach the groundwater. There we dig a waterhole to collect all of the London rain that falls in the area of the pavilion. In that way we incorporate an otherwise invisible aspect of reality in the park – the water under the ground – into our pavilion. As we dig down into the earth to reach the groundwater, we encounter a diversity of constructed realities such as telephone cables, remains of former foundations or backfills. Like a team of archaeologists, we identify these physical fragments as remains of the eleven pavilions built between 2000 and 2011. Their shape varies: circular, long and narrow, dot shaped and also large, constructed hollows that have been filled in. These remnants testify to the existence of the former pavilions and their more or less invasive intervention in the natural environment of the park.

All of these traces of former pavilions are now revealed and reconstructed. The former foundations and footprints form a jumble of convoluted lines, like a sewing pattern. A distinctive landscape emerges which is unlike anything we could have invented; its form and shape are actually a serendipitous gift. The plastic reality of this landscape is the perfect place to sit, stand, lie down or just look and be awed. In other words, the ideal environment for continuing to do is what visitors have been doing in the Serpentine Gallery Pavilions over the past eleven years. The pavilion's interior is clad in cork – a natural material with great haptic and olfactory qualities and the versatility to be carved, cut, shaped and formed.

On the foundations of each single pavilion, we extrude a new structure (supports, walls, slices) as loadbearing elements for the roof of our pavilion – eleven supports all told, plus our own column that we can place at will, like a wild card. The roof resembles that of an archaeological site. It floats a few feet above the grass, so that everyone visiting can see the water on it, its surface reflecting the infinitely varied, atmospheric skies of London. For special events, the water can be drained off the roof as from a bathtub, from whence it flows back into the waterhole, the deepest point in the pavilion landscape. The dry roof can then be used as a dance floor or simply as a platform suspended above the park. Herzog&de Meuron+Ai Weiwei

项目名称：Serpentine Gallery Pavilion 2012
地点：Kensington Gardens, London, UK
建筑师：Herzog & de Meuron, Ai Weiwei
项目团队：Ben Duckworth(Associate, Project Director), Christoph Zeller(Project Manager), Liam Ashmore, Martin Eriksson, Mai Komuro, Martin Nässén, John O'Mara(Associate), Wim Walschap(Associate), E-Shyh Wong, Inserk Yang
施工时间：2012.4.10—2012.5.25
开业日期：2012.6.1—2012.10.14
摄影师：©Julien Lanoo-p.14~15, p.18~19, p.20
©Roland Halbe-p.21

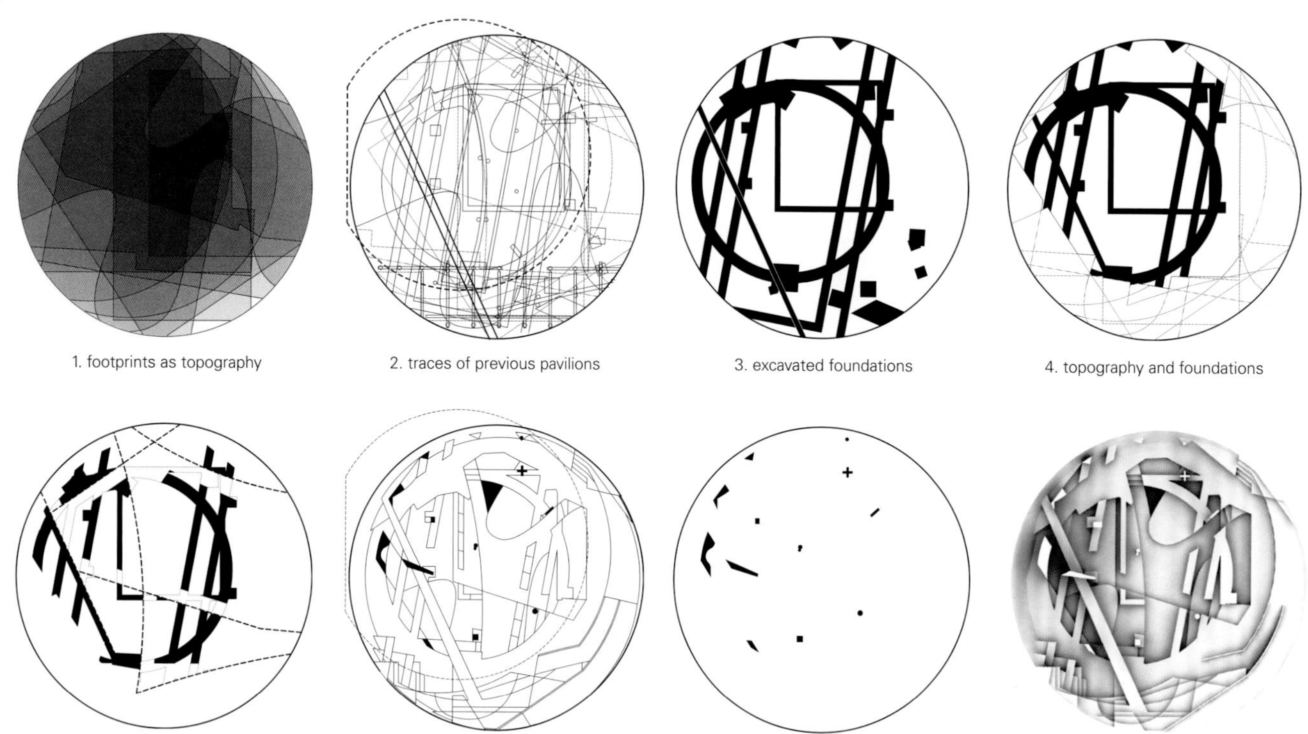
1. footprints as topography
2. traces of previous pavilions
3. excavated foundations
4. topography and foundations
5. cuts for circulation
6. extrusion of fragments
7. twelve specific columns
8. landscape

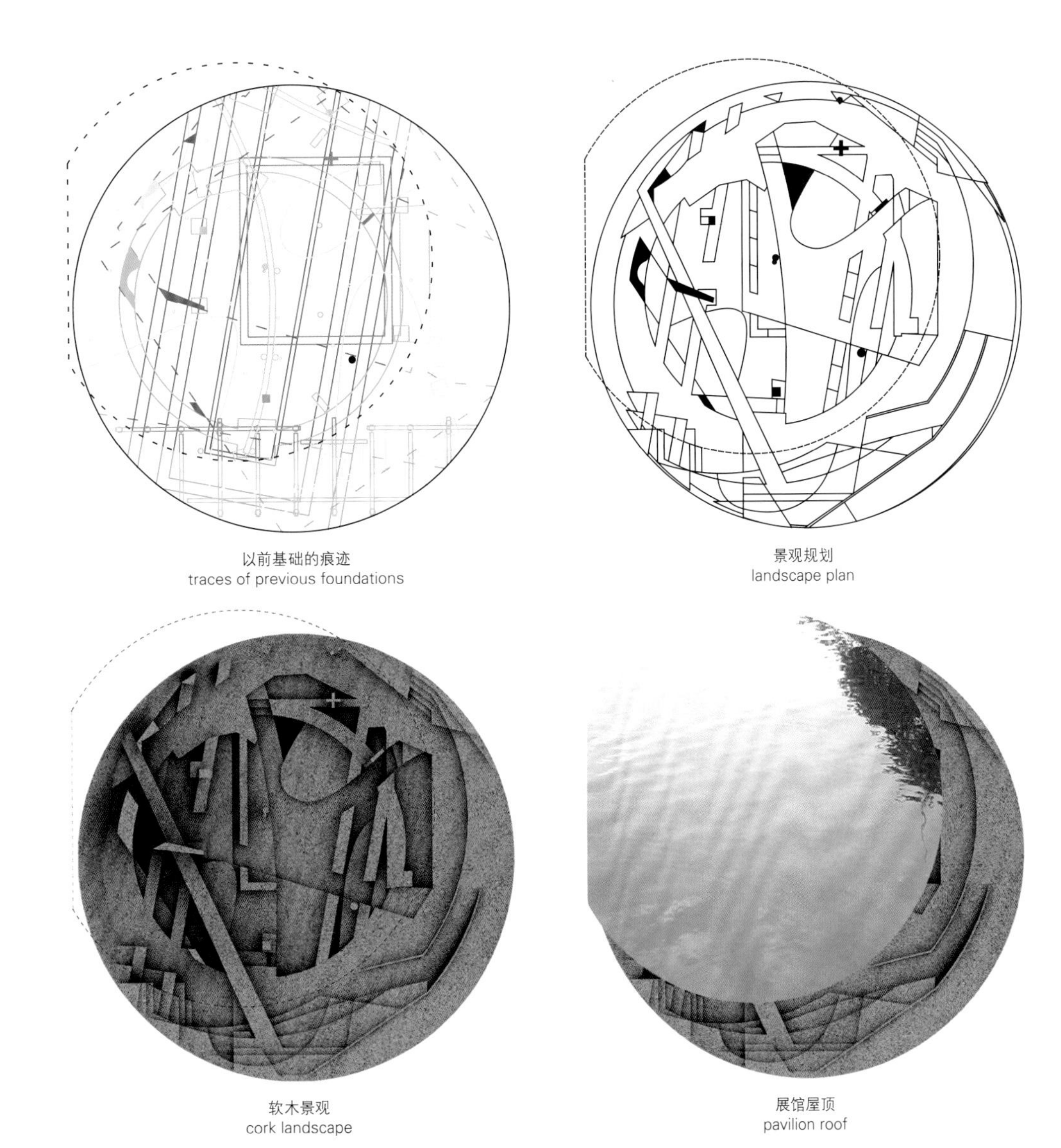

以前基础的痕迹
traces of previous foundations

景观规划
landscape plan

软木景观
cork landscape

展馆屋顶
pavilion roof

新医疗建筑

New Hospital(ity)

在过去的十年间，医院和医疗部门的运营模式发生了翻天覆地的变化。在许多国家，尤其是欧洲，原有的国立医院已变得更具竞争力、更加专业化，同时民营化的趋势也逐渐兴盛开来。对于大多数医院来说，治愈患者已不再是它们追求的唯一目标，客户的（空间）体验已变得与前者同样重要。对于其他医院来说，它们希望拥有能够快速运转的高效空间和程序，使患者的治疗过程顺利完成，并拥有良好的就诊感受，同时保证他们在医院的停留时间短、花费少。在其他地方，医院希望营造出舒适的环境，能够改变医院在重组空间方面的想法，同时让患者越来越欣赏这种更人性化的治疗方式，并减少他们的焦虑感。另外，几乎所有新型医院的设计都担负着减少二氧化碳排放量的责任，实现长期可持续发展的目标策略。所有的这些原因，不只影响着医院建筑本身，同时也影响着其所处景观和公共领域中周围文脉的设计。建筑的设计方案满足了所有医学界中流行的新趋势，设计师在新型医院中重新找回了古老的拉丁人设计医院的本质：医院环境舒适，氛围友好热情。

In the course of the last decade, the way in which hospitals and the medical sector operate has gone through a dramatic change. In many countries, especially in Europe, hospitals that were traditionally state-funded have become more competitive and are more specialized as privatization slowly takes place. For many, the treatment of the patient is no longer the only focus, but the (spatial) experience of the customer is becoming just as important. For other hospitals, it is to their interest of having an efficient space and procedures for a more rapid "turn-over", keeping the patient's treatment and experience smooth but period of stay, and costs, down. In other places, it is the medical interest of having a "well-being environment" that has changed the hospital's attitude in re-organizing their spaces as patients appreciate more and more the human approach that heals and reduces anxiety. In addition, almost all new hospital designs adopt the global responsibility and strategies of reducing carbon emissions and long-term sustainability. All of these reasons not only influence the architecture in hospitals itself but also the design for its surrounding context in landscape and the public realm. Architecture responds to all of these new tendencies in the medical world as designers rediscover in new hospitals the old Latin essence: hospitalis being hospitable, friendly and welcoming.

治愈未来
Healing the Future

全新模式

即使在当前这个经济低迷的时期，在科技日益先进、老龄化人口日益增加的大背景下，政府、国家和保险公司仍面临着医疗成本不断上升的巨大压力。医院自然也不例外。许多规模庞大且实行官僚制度的英国公立医院和综合医院已经累积了太多债务而无法支撑下去，私人公司已经开始接手管理。[1]商务专家曾提出、一些研究也表明：医院是可以自给自足的，前提是如果它们不接受那些不可能完成的任务（提高每个人的健康水平、保持低成本运营），而是将医院分为不同的解决方案小组[2]。实现医院自给自足可以通过以下几方面：建造建筑物独有的设施或"医院中的医院"，在这里，会计核算是彼此独立的，但治疗患者的过程却是完整且专注的。[3]这与其他类型的新商业模式形成了一系列全新的专业医院或专科医院，例如由赫尔佐格和德·梅隆建筑事务所设计的苏黎世儿童医院，抑或像由Pinearq设计的、位于巴塞罗那的格拉诺勒斯医院的门诊大楼这样的扩建项目。这种差别使得新型医院的设计与人们对传统医院的刻板印象相距甚远。在儿童医院的设计案例中，凉亭式的结构非常适合儿童使用。而格拉诺勒斯医院尊重历史设施，通过现代却中性均匀的设计方案力求将建筑与其融合在一起。这两个项目都打破了以往由于传统医院多重标准和规章制度而造成的支离破碎的空间，形成了简洁、清晰、完整的建筑。

空间、服务、体验的水准

随着就医选择范围的不断扩展，各大医院之间开始争夺患者，这使得医院将管理的重心转向了通过服务和体验来增加医院的附加价值。治愈疾病固然很重要，然而患者在治疗过程中的体验、感受以及享受的服务也同样重要。这种观念上的变化在医院的设计中也有所体现，这不仅仅是一个专门的医疗机器，同时也是一件艺术作品、一个温暖的地方，这种变化对近期的医院项目产生了重大影响。建筑师将长长的走廊改造为较短的漫步小路，位于玻璃中庭下方。同时将幽闭恐怖症患者的候诊室改造成了明亮的"休息室"，这里覆盖有无线网络并安装了其他设施，给患者带来更贴心的候诊感受。意式咖啡吧和出售健康产品的精

New Model

Even before the current economic downturn, governments, states and insurance companies were being challenged by the increasing costs of healthcare in an ever more technologically advanced world and an aging population. Hospitals are no exception. Large in nature, and often bureaucratic, many public and general hospitals in the UK have been accumulating so much debt that private companies have been brought in to take over management.[1] Business experts and studies suggests that hospitals can "heal" themselves financially if they are not tasked by the impossible job to improve health for everybody well and keep costs low, but be divided into solution-shops[2]. This can be achieved by building distinct facilities or hospitals-within-hospitals where accounting can be separated while the process of treating a patient is integrated and focused.[3] This and other new business models have led to a number of new hospital projects either as specialized or specialist hospitals, such as the Zürich Children's Hospital by Herzog & de Meuron, or as extensions such as the outpatient building at Granollers Hospital, Barcelona designed by Pinearq. The distinction steers new hospital designs away from the traditional hospital stereotypes. In the case of the Children's Hospital, into a pavilion-style structures that are child-friendly. The Granollers Hospital respects the historical facilities and strives to integrate itself through a modernistic but neutral and homogeneous design. Both projects break down often-fragmented spaces, stemmed from the usual hospital's multitude of levels and disciplines, into simple, clear and integrated architecture.

Quality of Space, Service, and Experience

As options for hospital treatment start to grow, the competition between hospitals for patients has led management to turn the focus onto added value through service and experience. Not only are treating the diseases important but the experience and process of how one is treated and served is equally important. This change in perception, of that hospital designs can be more than a specialized medical logistic machine but also a work of art and a place of warmth, has certainly the most impact in recent hospital projects. Long corridors have been reduced to a shorter walk under glass atriums. Claustrophobic waiting rooms have turned into day-lit "lounges" with Wi-Fi and other facilities catered for a better waiting experience. Coffee machines are replaced by espresso

照片提供：C.F. Møller Architects (©Torben Eskerod)

阿克什胡斯大学医院，呈现出城镇一般的结构，拥有广场和开放式空间
Akershus University Hospital, providing a town-like structure with squares and open spaces

品店代替了原有的咖啡机。这栋全新的医院建筑迎合了这些新趋势，从而为患者提供必要的空间感受。位于挪威奥斯陆附近的阿克什胡斯大学医院具有结构良好的开放式交通流线，这所医院是由C.F. Møller建筑事务所设计的，从健康信息到综合医院再到门诊外科手术，建筑师将注意力放在了功能的自然发展上。这座巨大的建筑综合体拥有城镇一般的结构，里面安置的设施均以患者为中心。这栋建筑形成了一种定义明确的生活，具有健康向上的社交活动氛围，室内日光充足，并与环境联系在一起。该建筑事务所所强调的品质在挪威卑尔根的Haraldsplass医院项目中也有所体现，然而在这所医院中，建筑师通过将两种类型的公共区域环绕在两个大型中庭周围，而使私人区域和公共区域有着清晰的界限。公共区域中包含商店、咖啡厅和休息区，私人的开放空间则是为患者及其探病的亲友所预留的。

空间效率

对于服务和体验的重视已经在改善医院的设施和空间品质的过程中变得显而易见了，然而我们也应该密切关注另一方面：那就是提高治疗过程中后勤服务的效率以及尽量简化住院治疗程序。首先，建筑师需要对治疗过程中的后勤服务效率进行规划，通过减少距离实现更好的布局，减少楼梯数量，或者像巴塞罗那格拉诺勒斯医院一样将公共路线和技术路线区分开来，这会让工作人员和患者感觉轻松一些。一些技术也提高了医院的运营能力，改善来访人员的感受。由于能够更准确地估计出候诊时间，并且医护人员能针对患者的医疗记录进行更好地沟通和登记，这使得患者能够在就诊或候诊时更好地利用自己的时间，而不会焦急无聊地在传统的候诊室里等待。

一种治疗观点：景观与舒适的环境

提高病人的康复效率以及尽量简化住院治疗程序的主旨，与将“舒适的环境”融入医院当中的理念息息相关。改善环境能够创造出一个更好的治疗地点，这种想法由来已久。研究表明，日光能够减少焦虑和抑郁。身处大自然和花园之中，能够降低血压、减少压力、改善情绪。[4]尽管

bars and boutiques selling health products. The new hospital architecture caters to these new tendencies in order to deliver the necessary spatial experience required. These characters of open and well-structured circulation can be seen in projects such as Akershus University Hospital near Oslo (Norway), where C.F. Møller Architects focused on a natural progression of functions from health information to polyclinics to outpatient surgeries. The large complex is structured like a town where the facilities are centered on the patient. Here the architecture ensures a well-defined daily life with a healthy level of social contact, daylight, and connection to the environment. The same architects emphasize these qualities in the Haraldsplass Hospital project in Bergen (Norway) but defines the private and public with clarity by having two types of common areas surrounding two large atriums. A public area consists of shops, cafe, and a lounge area, while a private open space reserved for patients and theirs guests.

Efficiency of Space

While the emphasis on service and experience has been evident in its translation into a better quality of facilities and spaces in hospitals, another emphasis being closely looked at is the efficiency in the logistics of treatment and in minimizing hospitalization. The first, efficiency of logistics of treatment, is being addressed architecturally through better layouts in minimizing distances, less corridors or by efforts in differentiating public and technical routes such as that of Granollers Hospital in Barcelona. This helps the staff and patients alike in reducing fatigue. Technology also improves operations and logistics of a visitor's experience. As waiting time becomes better estimated, and patient's medical records better communicated and registered, patients can make better use of their time during their visit or pleasantly wait rather than feeling bored or anxious in the traditional waiting room.

A View to Heal: Landscape and Well-being environment.

The aim, efficiency to recover and minimize hospitalization, comes hand in hand with the concept of integrating "well-being environment" into hospitals. The idea that improving the environment creates a better place to heal is not new. Research has showed that daylight reduces anxiety and depression. Physical and visual exposure to nature and gardens reduces blood pressure, stress, and improves moods.[4] The importance of connecting green and nature into the new hospital is a recurring theme, however archi-

Rigshospital医院内外绿色景观的融合
the integration of green both inside and outside in Rigshospital

将绿色和大自然融入医院的重要性是一个永恒的主题，然而建筑师们却运用了不同的设计方案将周围环境和景观融入在内。

高效的后勤系统和舒适的环境这一理念是哥本哈根Rigshospital医院扩建项目的设计基础，在该项目中，建筑师将其设计方案塑造为一系列V形建筑，以求最大限度地获取日光和附近公园中的绿色自然景观。3XN建筑事务所提出的设计方案既优雅又高效，一条"快速通道"将所有的侧翼建筑都连接在一起，以形成高效省时的后勤系统。

在"构造"自然、引入自然光线方面，庭院的用途显而易见，同时它还可以过滤噪音、屏蔽周围的环境。由赫尔佐格和德·梅隆建筑事务所设计的苏黎世儿童医院运用小巧且私密的设计方法打造出几何形庭院，并尝试通过外部的个别开口将建筑和自然与乡村的环境交织在一起，以便于日光可以照射到较低的建筑物中。木质立面反映出了这片区域中的自然风光。在天井周围构造这样的形式是为了让患者及其家属可以随意在其中漫步。

由MGM建筑事务所设计的先进医疗技术医院研究所位于塞维利亚，其设计方案是将其中一个庭院作为主要的交通流线区域和入口，其余的庭院作为沟通"节点"将不同的大厅与不同的部门连在一起。然而庭院内部却没有包含景观，但是这种紧凑高效的设计方案将重点放在了自然光线的引入方面，建筑师力求让所有重要的功能区域都能感受到自然光线，尤其是住院患者的病房，希望让他们拥有舒适的体验。连续不断的有孔立面内外都设有大型玻璃窗，将不同的功能和谐统一起来，同时淡化透明与坚固之间的界限。

西班牙穆尔西亚地区气候干燥，因此Casa Sólo建筑事务所在设计新圣卢西亚大学综合医院时，并没有将设计重点放在绿色植物上，而是更加注重运动、休闲和商业活动方面。这所大型医院更加注重预防方面的作用，力求使社区与健康空间以及像游泳池、屋顶花园和庭院这样的设施融为一体。

优秀的医院设计理念可以缩短患者的康复时间，这一点是显而易见的，建筑师和用户负责阐释"是什么决定了舒适的环境"这一话题。以下是患者如何感知"舒适的周边环境"的一个实例：在荷兰，女性有

tects have used different design methodologies to embrace surroundings and landscape.
The concept of efficiency in logistics and well-being is the basis that has sculpted the design scheme of Rigshospital Expansion in Copenhagen into a series of V-structures in plan to maximize daylight and views to green and nature in adjacent parks in the urban environment. An elegant and effective solution designed by 3XN, a "fast track" connects all wings for efficient timesaving logistics.
The uses of courtyards are apparent in "framing" nature and bringing in light while filtering out noise and the immediate environment. Herzog & de Meuron's Zürich Children's Hospital employs geometric-shaped courtyards in a small and intimate manner as it tries to weave architecture and nature through occasional openings outside to the rural setting to welcome daylight into the low building. Its wooden facade reflects the natural surrounding of the area. The form is arranged around the patios so that patients and their family can move around as freely as possible.
Located in Seville, the Institute of Advanced Medical Techniques Hospital designed by MGM Arquitectos, employs one of its courtyards as main circulation and entrance, while the rest as communication "nodes" to connect various lobbies to different departments. The courtyards contain no landscape however, but this compact and efficient design concentrates on bringing in natural daylight to all of the important functions notably the in-patient wards for its well-being experience. A continuous perforated facade plays with glass planes inside and outside, harmonizing different functions while blurring the difference between transparency and solidity.
In drier climates such as Murcia (Spain), Casa Sólo Arquitectos focuses less on greenery but on sports, leisure, and commercial activities at the New Santa Lucia University General Hospital. This large hospital takes a more preventive role as it strives to integrate the community into its health-related spaces and facilities such as swimming pools, rooftop garden, and courtyards.
While the idea of good hospital design to reduce patient's recovery time is clear, the topic of what defines a well-being environment is open for interpretation by the architect and the user. As an example of how patients perceive "well-being surroundings", women are given the choice in the Netherlands to give birth where they feel most at ease. Some feel less anxiety to give birth at home with midwives, while others prefer to give birth in a fully

圣卢西亚大学综合医院编织状的建筑外围护结构，与周围的山地景观完美融合
Santa Lucia University General Hospital with textile architectural envelope, shaped to get along with the hilly surrounding landscape

权利选择在她们感觉舒适的地方生孩子。一些人认为自己在助产士的帮助下，在家生产不会感到那么焦虑，而另一些人则更愿意在设施完备的医院里生下自己的宝宝。尽管如此，还有一些女性愿意留在"母婴酒店"生产，因为这里的环境舒适温馨，附近还备有相应的医疗设施及医护人员。因此形成的不同感知也就解释了为什么在这座新型医院里需要形成不同类型的舒适环境。

Rafael de La-Hoz建筑事务所设计的胡安卡洛斯国王医院，其设计重点既不在周围的文脉中也不在自然之中，而是通过高效、明亮和宁静将重点放在了将"市民转变为顾客"这一方面。医院位于马德里的Móstoles地区，这所医院将舒适因素的重点放在了减少太阳能得热和注重私密性上，并结合了住宅建筑的特色。传统的病房利用窗帘来隔开共用的病床，而该设计方案却截然相反，环绕在两个玻璃天井周围的房间非常私密宁静，与手术室和下方的其他设施近在咫尺。而透过圆形玻璃幕墙看到的外部景色是变形的，一些人对此的理解是这里的建筑试图将患者的注意力转向内部，同时设计师想要打造一个人性化的治愈场所。

可持续发展

除了营造舒适的环境、高效的后勤系统以及超一流的服务之外，新型医院及其建筑师也担负着更深远的责任：可持续发展。由于医院的规模庞大，同时又需要昼夜不停地运转，再加上使用的设备、通风系统、供暖设施和其他设施对能源的需求，因而可持续发展以及减少二氧化碳的排放量也就成了一个重要问题。在欧洲，建筑物二氧化碳的排放量占总排放量的40%，医疗行业建筑的排放量在其中一直都排名前列。[5]每一份设计概要的可持续发展评估都推动着医疗设施的改革不断向前迈进，以减少包括能源、水、废物在内的资源消耗。然而可持续发展在不同的环境中有着不同的含义，并没有明确的界定。这所新型医院的建筑师们采用了始终如一但更加宽泛的可持续发展设计方案。设计中往往

equipped hospital environment. Still, other women prefer to stay in "birth hotels", where the environment is made out to be cozy and homely but with the medical facilities and staff near by. This difference of perception explains the variety of well-being environments needed in the new hospital.

The Rey Juan Carlos Hospital designed by Rafael de La-Hoz, focuses neither on surrounding context nor nature, but on turning the "citizen into a customer" through efficiency, light, and silence. Located in Móstoles, Madrid, the well-being factor of this hospital focuses on reducing solar gain, privacy, and incorporating residential architecture as part of the experience. As opposed to traditional wards, which may have shared beds separated by curtains, the rooms surrounding the two glass atria volumes are private and peaceful, effectively close to operating rooms and other facilities below. However, the view to the outside is distorted through a bubble glass facade. One gets the idea that the architecture here wants to turn its patients inwards, as designers attempt to create a human-approach, therapeutic healing machine.

Sustainability

Alongside creating well-being environments, efficient logistics, and outstanding service, new hospitals and its architects face a further responsibility: sustainability. Due to a hospital's scale, around-the-clock operation, its demand for energy through usage of equipment, ventilation, heating and other utilities, sustainability and reducing carbon emissions are a big topic. In Europe, buildings claim 40% of CO_2 emissions; healthcare sector is a leader in the market.[5] Sustainability assessments of every design brief drive forward healthcare facilities to reduce their resource consumption including energy, water, waste. Yet, sustainability has often very different meanings in different context, and can create a lack of clarity. Architects of the new hospital have adopted a consistent but a wide view of sustainability. Designs often address sustainability through immediate architectural solutions such as solar protection or insulation factors but hospital designers must consider sustainability in the lifetime of the buildings and outside of just energy consumption. Other types of sustainability measures include developing a whole life costing approach, creating therapeutic environments, and flexible estate planning.[6] Estate planning was in the past overlooked, as many hospitals did not effectively masterplan their estate resulting in an unhealthy

胡安卡洛斯国王医院的玻璃中庭，这里拥有与患者病房相似的宁静氛围
glass atriums of Rey Juan Carlos Hospital with peaceful atmosphere close to the patient's room

通过直接的建筑解决方案来阐述可持续发展，例如遮阳系统或保温装置，但是医院的设计师必须在建筑物的使用寿命中以及在能源消耗之外考虑可持续发展问题。其他类型的可持续发展措施包括开发终生成本方案、营造治疗环境以及灵活的房地产规划氛围。[6]过去人们常常忽略房地产规划，因为许多医院并没有对其房地产进行有效的总体规划，这导致了建筑物在高密度的扩建过程中形成了不健康的组团。城市设计与优秀的景观建筑相结合、建筑场地的融合及保存也在新医院项目中巩固了可持续发展策略。

C.F. Møller建筑事务所设计的阿克什胡斯大学医院采用了可持续性的设计方案，实现了高水准的成就。在其他设计方案中，最令人印象深刻的是通过地热能源提供85%的热量和40%的能源消耗总量。医院寄希望于可再生能源设备，这样剩余热量就能够存储在地下，比原有医院减少了50%的二氧化碳排放量。如果考虑到医院的能源消耗量相当于1300个家庭的总耗量，那么这一比例是巨大的。建筑师在设计Haraldsplass医院时也采用了类似的设计方案，然而不同的是设计中多了一些简洁的特性，重新利用了现有医院中的废热。将病房与新式通风设备结合在一起，使前者也可以像被动式建筑一样得到设计。

3XN建筑事务所在设计Rigshospital医院扩建项目时采用了可持续发展方案，室内外均呈现出完整的绿墙和花园，这有助于营造出舒适的环境，同时也能最大限度地减少对环境的负面影响。

Rafael de La-Hoz建筑事务所在设计胡安卡洛斯国王医院项目中融入了许多可再生的能源技术，但是人们能感知到的只有通过屋顶射入的自然光线以及通风系统。

从适应新的商业模式，到提升空间品质和高效的后勤系统，再到增加可持续发展的责任感，这些众多的全新参数和外力影响造就了今天的医院。这些挑战需要新的资源，同时也需要全新的思考方式和设计方式。它为改善和营造更舒适的治疗空间提供了机遇。随着技术和医学界不断地自我更新，当今的建筑师也需要重新思考未来的医院走向。

clustering of high-density extensions of buildings. Urban design in conjunction with good landscape architecture, site integration and preservation can underpin sustainability in the new hospital projects.

C.F. Møller Architects' Akershus University Hospital takes sustainable design to a high level of achievement. Among other interventions, the most impressive is that 85% of the heat and 40% of total energy consumption are provided through geo-thermal energy. The hospital invested in a renewable energy plant where surplus heat can be stored underground, reducing CO_2 by over 50% compared to former hospitals. This is huge when consider that the hospital consumes the equivalent of 1300 homes. The architects' approach to their Haraldsplass Hospital project takes similar but more modest approach by reusing waste heat from the existing hospital. Combined with new ways of ventilation, the ward can be considered similar to a Passivhaus.

3XN's solution to sustainability in Rigshospital Expansion is integral green walls and gardens both inside and outside, contributing to atmosphere of well-being while minimizing negative environmental impact.

At the Rey Juan Carlos Hospital, Rafael de La-Hoz incorporated many renewable energy technologies but most perceptible is the natural light and ventilation provided through the roof.

There are many new parameters and forces that shape the hospital today, from adapting to new business models, to improving spatial qualities and efficient logistics, to stepping up to the new responsibility of sustainability. These challenges require new resources but also a new way of thinking and design. It opens up opportunities to improve and create a more hospitable place for healing. As technologies and the medical world renew themselves, architects today require a rethink of the hospital of the future. Andrew Tang

1. http://www.guardian.co.uk/uk/2011/nov/10/private-nhs-hospital-accident
2. http://www.forbes.com/2009/03/30/hospitals-healthcare-disruption-leadership-clayton-christensen-strategy-innovation.html
3. http://www.forbes.com/2009/03/30/hospitals-healthcare-disruption-leadership-clayton-christensen-strategy-innovation.html
4. "The psychological and social needs of patients", British Medical Association, January 7, 2011, retrieved on March 14, 2011
5. Jonathan Erskine & Joram Nauta, Lead Market Initiative–*Low Carbon Buildings in the Healthcare Service Sector*, DuCHA workshop, June 9, 2010 (http://ducha10.ducha.nl)
6. Phil Nedin, *Designing Healthcare Facilities through a Sustainable Model; New Perspectives*, DuCHA Workshop, June 8, 2010 (http://ducha10.ducha.nl)

Haraldsplass医院

C.F. Møller Architects

评审团描述C.F.Møller建筑师事务所设计的获奖项目是一个完全新型的医院，它是位于挪威卑尔根Haraldsplass医院的新病房区建筑，占地10 000m²。医院传统的走廊现在被开放的公共区域和高效的后勤设施所替代。Haraldsplass医院建于1986年，有大概184个病床。新建筑将占地10 000m²，并分三个楼层增加108个病床。同时还会修建新的地下停车设备，可容纳近400个车位。

新建筑将位于Ulriken山脉的脚下，前方就是Møllendalselven河。评审团曾这样赞赏该项目在医院设计上的创新手法："在这里患者至上，他们被安置在可以看到多种美丽景色的区域里。吸引人眼球的绘画作品也同样带给病人和职工以全新的体验，同时病人家属也会被妥善安置在一个幽雅的、细致入微的到达区。"

该建筑完全不同于传统的医院建筑结构，没有长长的走廊。病房区位于两个大型的有顶的中庭周围，这里可以为两种不同的公共区域提供背景：一个是设有接待处、咖啡厅、商店和休息区的公共到达区和一个仅为病人及其亲友准备的更私人的空间。中庭可以使光线照射进建筑内，水池中葱翠的草本和竹类植物加上花坛中的小草、花卉及藤类植物都有助于提供良好的室内环境。由于新建筑环绕着Møllendalselven河，且立面倾斜了一定的角度，因此所有的病人都可以欣赏到山谷和城市的风景。

该项目同时也凸显了其生态友好性，除了别的原因，还因为其立面规格小于建筑面积。通过采用新式通风设备并重新使用原有医院的废弃热量，新式病房区可以达到"被动房屋"的标准。

该项目预计于2014年开始施工，于2015年竣工。

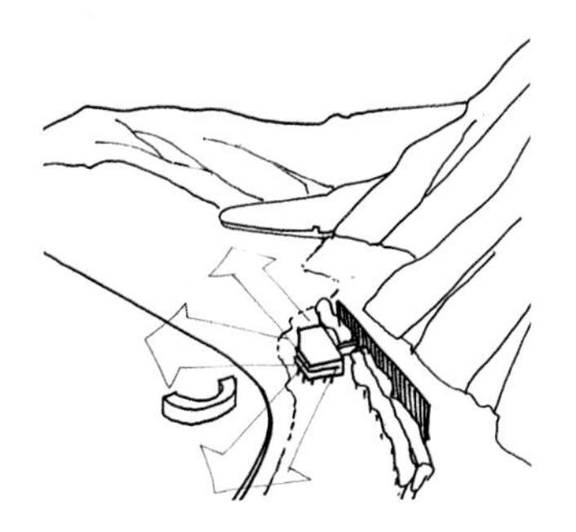

situation at the foot of the mountain

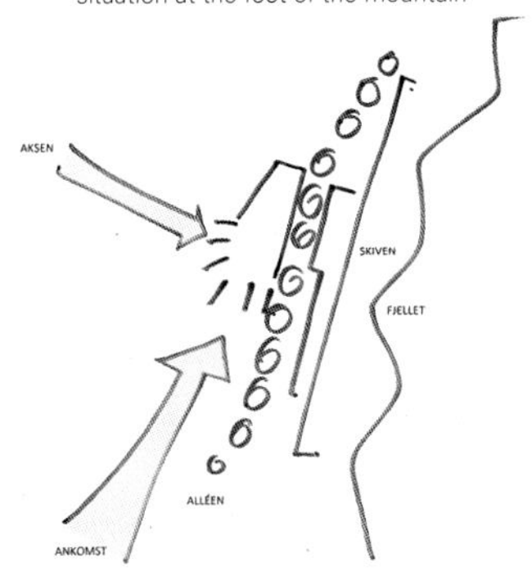

arrival, avenue, slab, axis & mountain

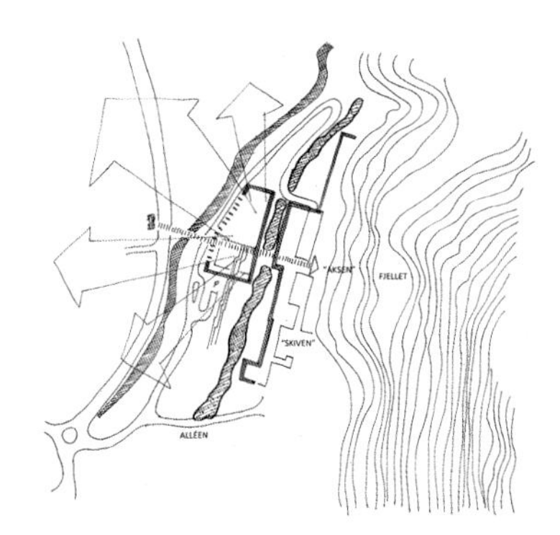

main concept, situation and orientation of the building: the new wing opens toward the city, with the mountains as backdrop

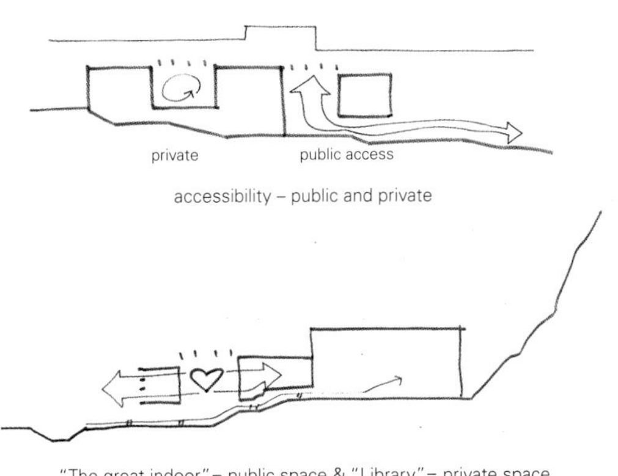

accessibility – public and private

"The great indoor" = public space & "Library" = private space

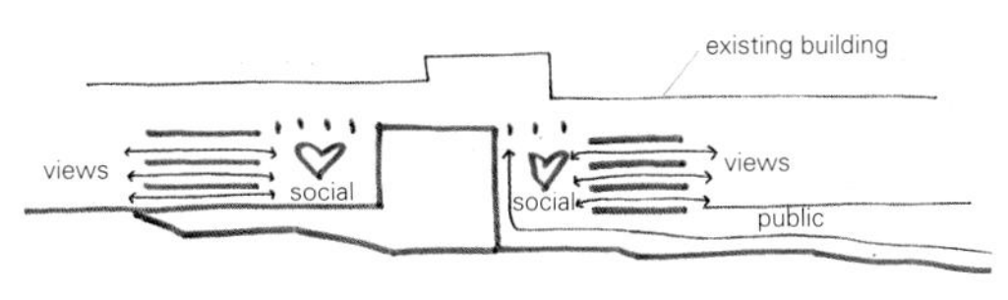

The organization of volumes and main functions (relationship to room, existing buildings)

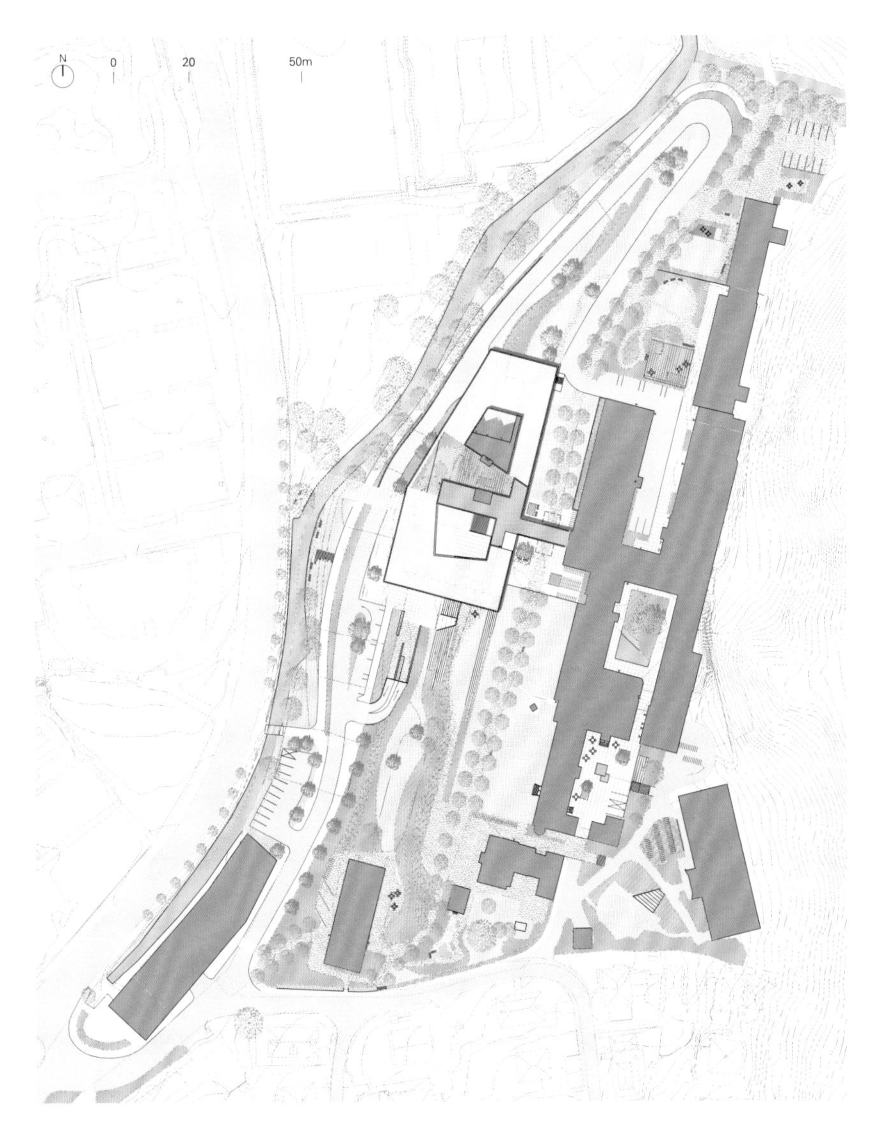

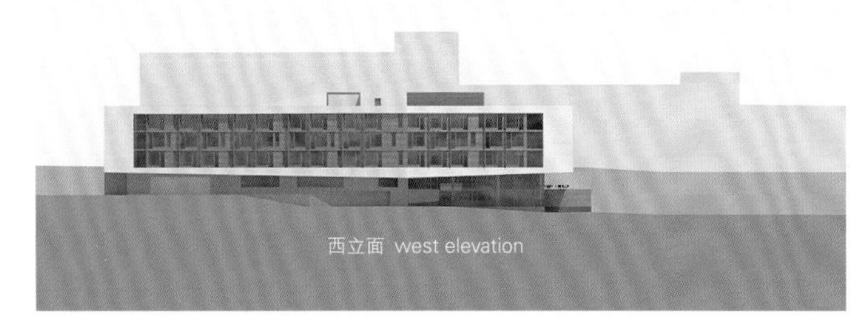

西立面 west elevation

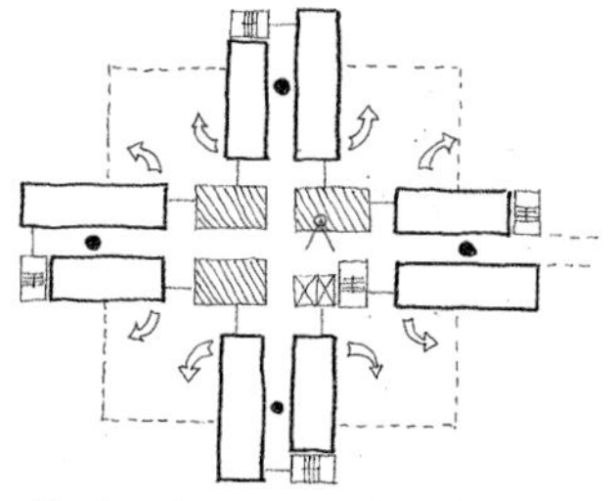

typical "cruciform" bed ward layout – not suitable for the location and brief

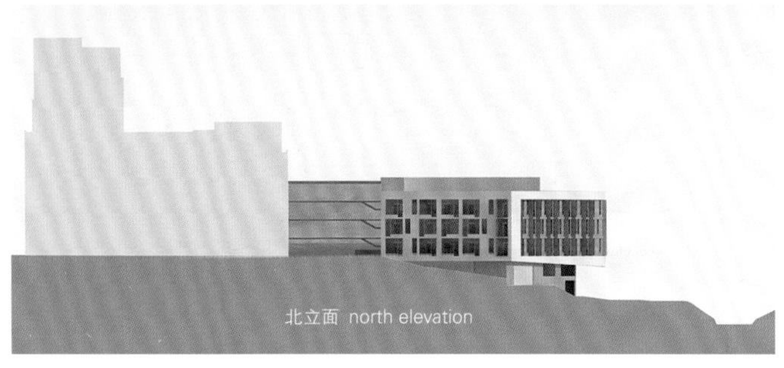

北立面 north elevation

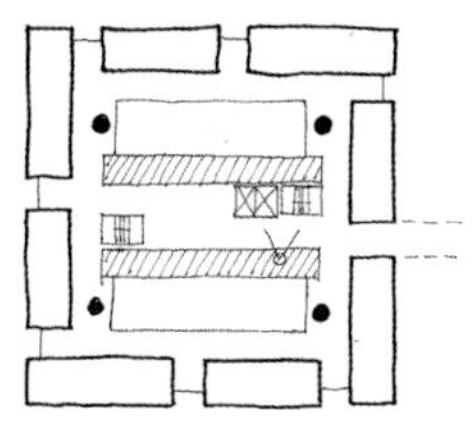

alternative "atrium-type" bed ward layout – achieves a compact building volume, fewer stair cores, vertical connections and eliminates traditional corridors.

南立面 south elevation

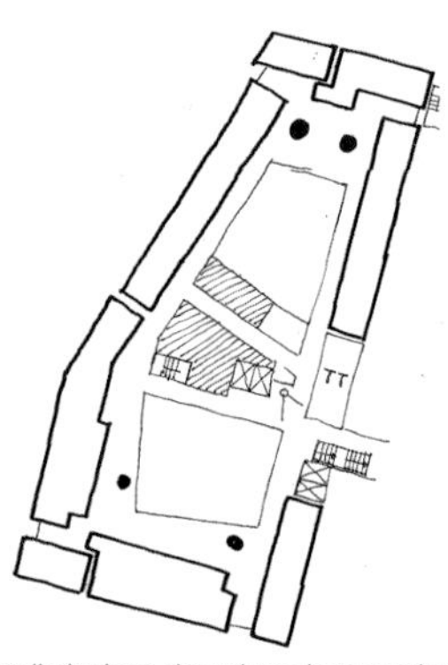

to align the layout with the plot limitations, the volume is tapered to fit the existing buildings.

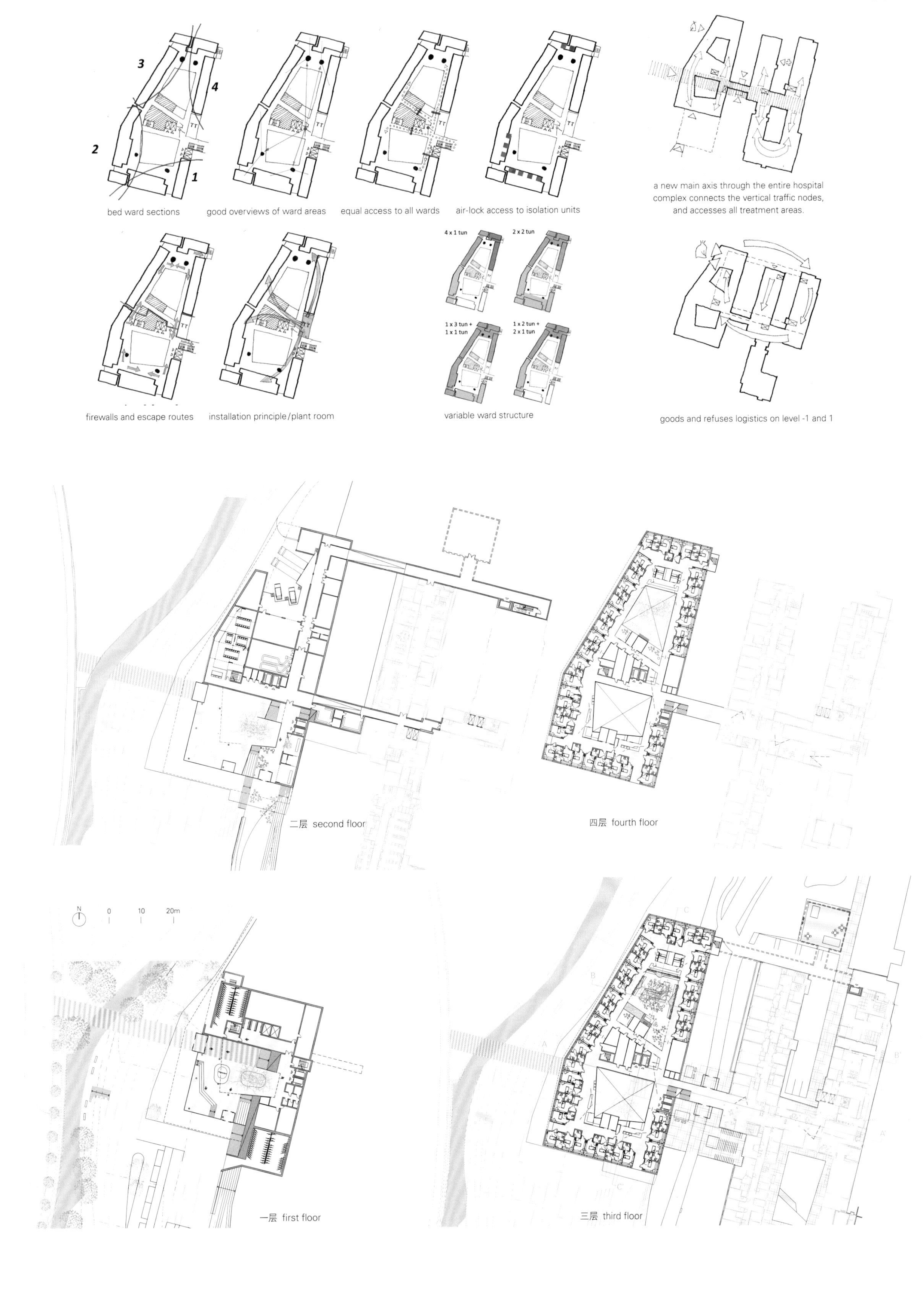
3
4
2
1
bed ward sections
good overviews of ward areas
equal access to all wards
air-lock access to isolation units
a new main axis through the entire hospital complex connects the vertical traffic nodes, and accesses all treatment areas.
firewalls and escape routes
installation principle/plant room
4 x 1 tun
2 x 2 tun
1 x 3 tun + 1 x 1 tun
1 x 2 tun + 2 x 1 tun
variable ward structure
goods and refuses logistics on level -1 and 1
二层 second floor
四层 fourth floor
N
0
10
20m
一层 first floor
三层 third floor

Haraldsplass Hospital

A whole new kind of hospital – that is how the jury described the winning proposal by C.F. Møller Architects for a new 10,000m² ward building for Haraldsplass Hospital in Bergen, Norway.

Haraldsplass Hospital was built in 1986 and has approximately 184 beds. The new building will cover 10,000m² and give the hospital a further 108 beds on three stories. There will also be new underground parking facilities for approximately 400 cars.

The new building will lie at the foot of the Ulriken mountain, with the river Møllendalselven in front.

In stark contrast to traditional hospital buildings, there are no long corridors. The wards are located around two large covered atriums, which provide the setting for two different kinds of common areas: a public arrivals area with a reception, cafe, shop and seating area, and a more private space for patients and their guests only.

The atriums draw daylight into the building, where lush vegetation with bamboo plants in water pools and a bed of grass, flowers and creeping plants help to ensure a good indoor climate. All patients will have access to views of the valley and the city, as the new building follows the course of the Møllendalselven river, with an angled facade.

The project has also been highlighted as being very eco-friendly, amongst other reasons because the facade size is relatively small to the gross area. By taking new approaches to ventilation and reusing waste heat from the existing hospital, the new ward can achieve "passive house" standard.

Work on the project is expected to commence in 2014 and be completed in 2015. C.F. Møller Architects

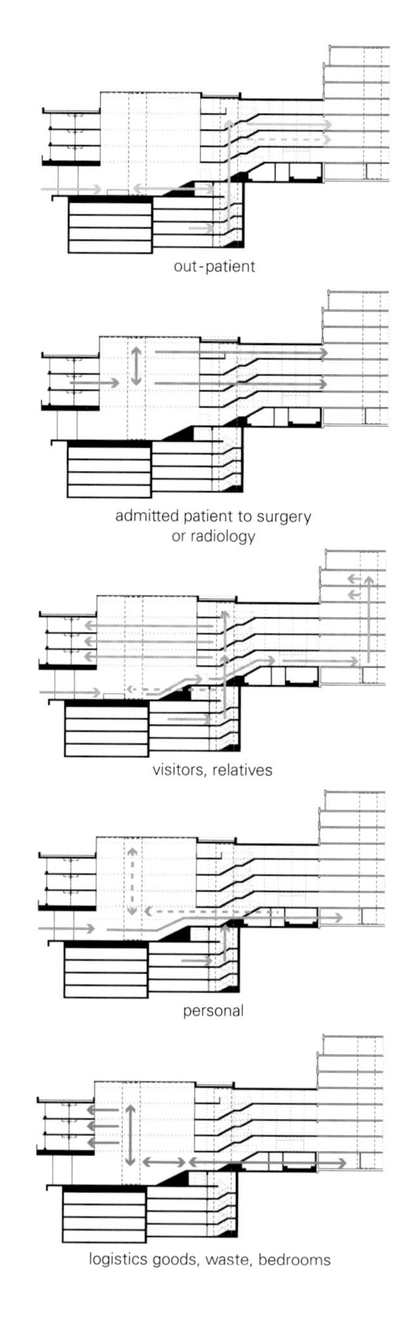

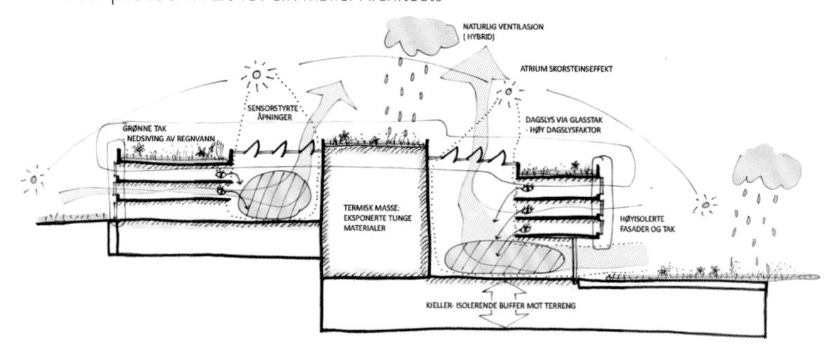

项目名称：Haraldsplass Hospital
地点：Bergen, Norway
建筑师：C.F. Møller Architects
工程师：Norconsult (energy consumption)
景观建筑师：Asplan Viak
用途：hospital
总楼面面积：10,000m²/108 beds (10,000m² parking/400 parking spaces)
设计时间：2011
施工时间：2014—2015 (expected)

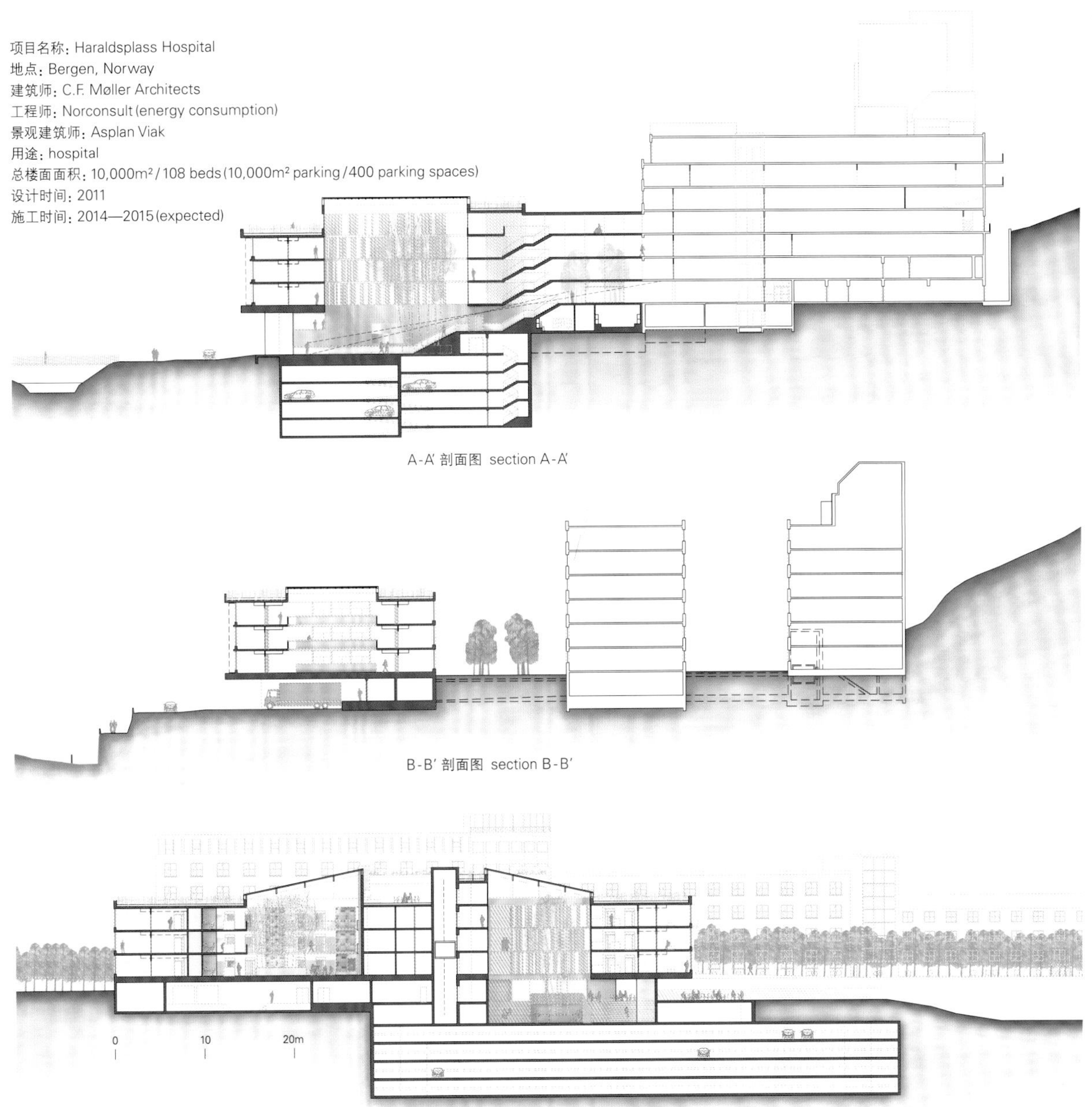

A-A' 剖面图 section A-A'

B-B' 剖面图 section B-B'

C-C' 剖面图 section C-C'

Rigshospital医院扩建

3XN

一支由3XN建筑事务所、Aarhus建筑师事务所、Nickl及合伙人建筑事务所、Grontmij与Kirstine Jensen工作室所组成的团队在对哥本哈根最重要的医院Rigshospital进行扩建的竞赛中获胜，该项目预计于2017年年初竣工。

这座地处哥本哈根市中心的医院将扩建76 000m²，其获胜的设计提案保证了建筑高效、省时的运营方式，同时日光、绿色空间以及周边公园的景色也为患者、员工和探视者带来了良好体验。

该建筑物的形状以一系列的折叠V形结构和一条横向快速通道为特色。

该结构将对每天在病房内工作的医护人员产生积极的影响，会使服务人员更高效地在建筑内行走，更重要的是可以给患者带来更加清静的环境。折叠的V形结构形成了五座中庭，这里可以为人们提供舒适的休息区域，而且还有着清晰可辨的寻路方向标志。最后，该建筑的形状确保了室内能够光线充足，并且置身其中可以将公园与城市的景色一览无余。

可持续性的建筑解决方案、花园以及建筑内外的绿墙都是整个设计过程中重点考虑的因素。将内外绿色环境融合起来的设计理念会使医院的氛围有助于患者的康复，也为员工和来访者营造了一种积极的环境，同时还将负面的环境影响降至最低。

为了适应周围的城市空间，该建筑的规模从西北部向东南部逐渐缩小。这意味着新建筑将比原有医院综合体高出九层，而临街的部分则只高出了四层。这样一来，新建筑的高度就可与街道另一侧的传统多层住宅互为呼应。由玻璃和浅色天然石块搭建的立面呈现出一定角度，这样可以达到部分自我遮阳的效果，并且在视觉上也与该区域的现有建筑立面相协调。

该项目除了扩建医院之外，还包括开发该地区的总体规划、建造一座新的患者酒店以及一个多层停车场。

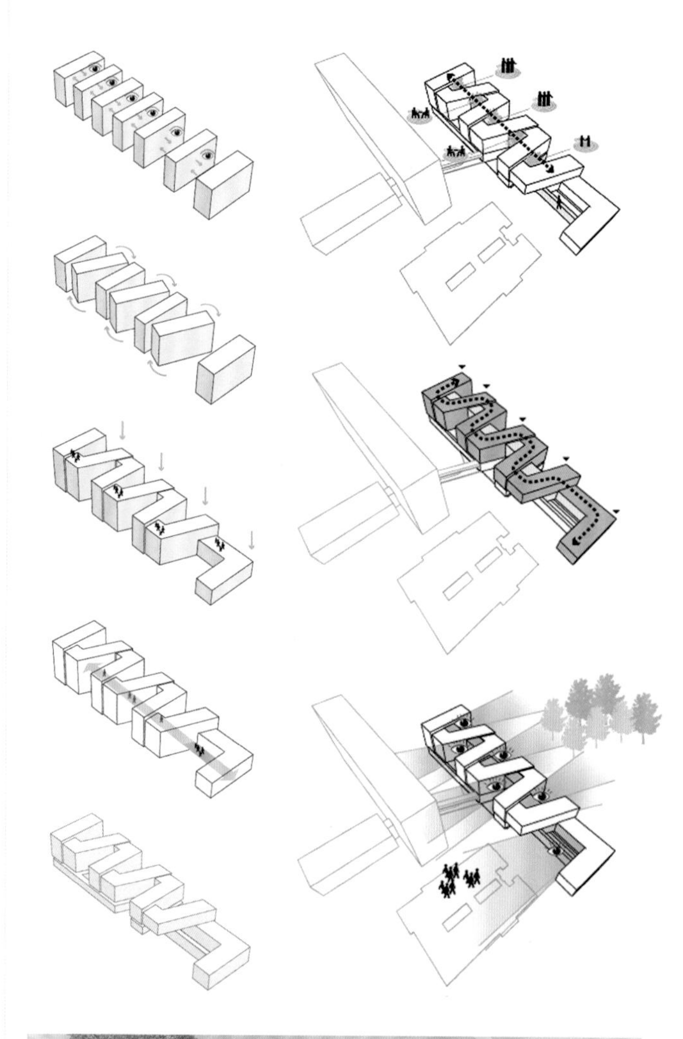

Rigshospital Expansion

A team of 3XN Architects, Aarhus Architects, Nickl & Partner Architechten, Grontmij and Kirstine Jensen Studio has won the prestigious competition for the expansion of Copenhagen's main hospital, Rigshospital, which is expected to be completed in early 2017. The winning proposal for the 76,000m² extension of Copenhagen's most centrally located hospital ensures efficient and time-saving logistics, while daylight, green spaces and views of the neighboring park contribute to the well-being of patients, staff and visitors.

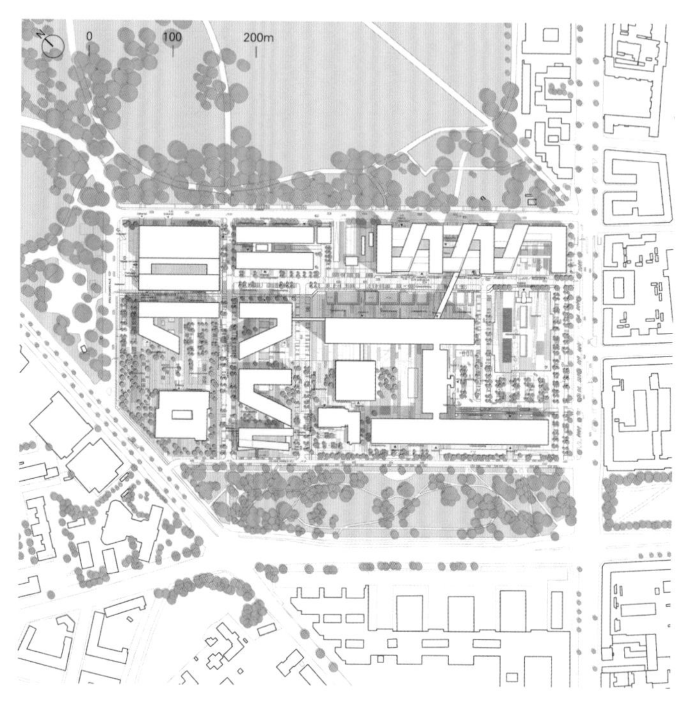

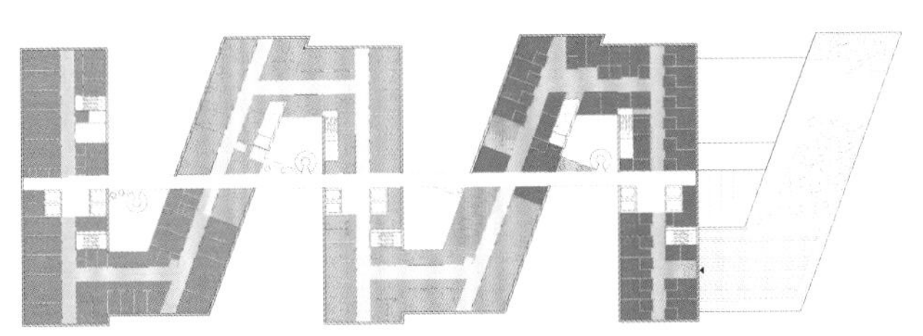

八层 eighth floor

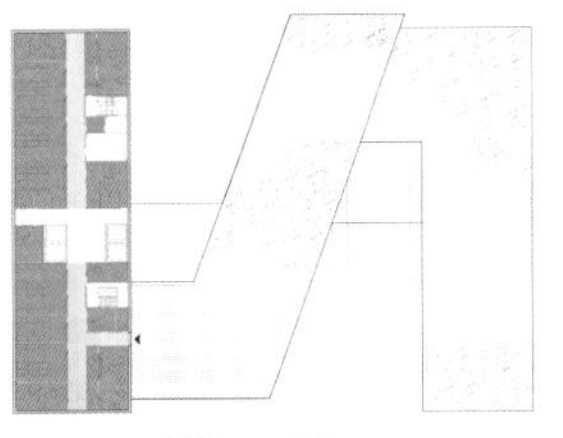

十层 tenth floor

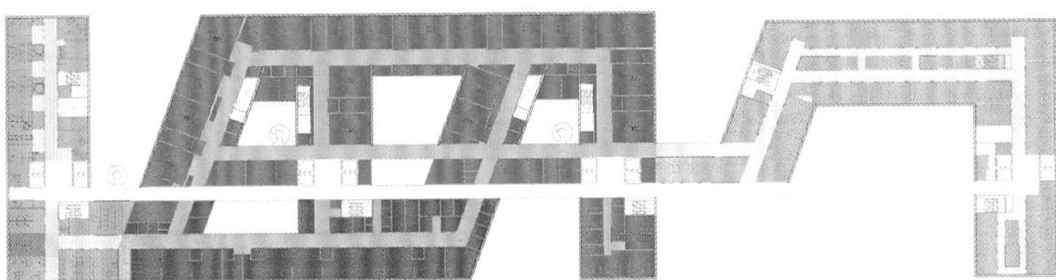

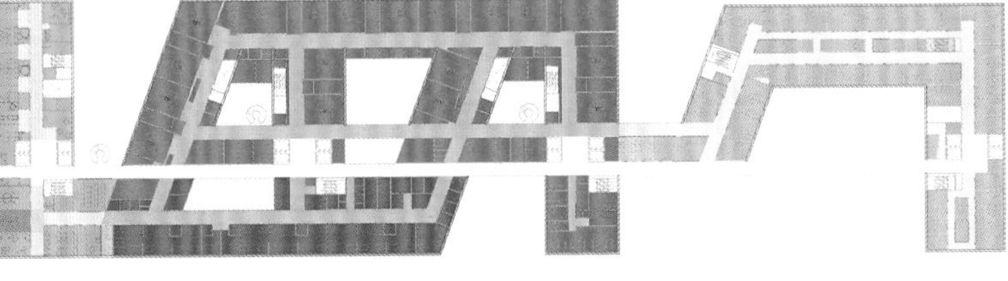

二层 second floor

项目名称：Rigshospitalet Expansion
地点：Blegdamsvej, Copenhagen, Denmark
建筑师：3XN
医院建筑方面的专家：Nickl & Partner Architekten AG
工程师：Grontmij
景观建筑师：Kristine Jensens Tegnestue
总顾问：Aarhus Arkitekterne
甲方：Rigshospitalet
材料：glass, Jura Gelb natural stone
总楼面面积：76,000m², patient hotel_7,400m², multi-storey car park_17,000m²)
竣工时间：2017(expected)

0 20 50m

一层 first floor

A-A′ 剖面图 section A-A′

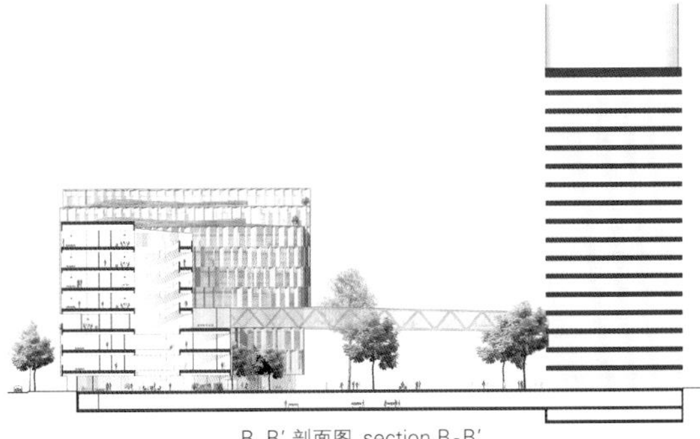

B-B′ 剖面图 section B-B′

The building's shape is characterized by a series of folded v-structures together with a transversal fast track.
The structure will have a positive impact on the everyday life of the staff in the wards: it will give the service personnel the opportunity to effectively move around in the building, and not least, it will give the hospital the opportunity to offer patients more calm and clarity. The folded v-structure provides space for five atriums that will serve both as comfortable lounge areas and points of orientation that make way-finding easy and logical. Finally, the shape of the building ensures an interior which is characterized by lots of daylight and views of the park and the city.
The integration of sustainable building solutions, gardens and green walls both inside and outside has been a key focus throughout the design process. The idea is that the integration of a green environment both inside and outside will contribute to an ambience that has a healing effect on patients and will also create a positive physical environment for staff and visitors. At the same time, these features will minimize negative environmental impacts.
To adapt to the surrounding urban space the building is scaled down from the northwest to the southeast. This means that the building has nine floors up against the existing hospital complex and just four floors towards the street. In this way, the building height matches the classic multi-storey houses on the other side of the street. The facade of glass and light natural stone is angled so that it is partially self shading and visually adapts to the area's existing building facades.
Besides the construction of the hospital extension the project includes the development of the area's master plan and the construction of a new patient hotel and multi-storey car park. 3XN

项目名称：Children's Hospital Zürich
地点：Zürich, Switzerland
建筑师：Jacques Herzog, Pierre de Meuron, Christine Binswanger(Partner in Charge)
项目经理：Mark Bähr
项目助理经理：Jason Frantzen
项目团队：Alexandria Algard, Alexander Franz, Ondrej Janku, Christoph Jantos, Johannes Kohnle, Blanca Bravo Reyes, Raúl Torres Martin, Mika Zacharias
结构工程师：ZPF Ingenieure AG, Basel
景观建筑师：August Künzel, Münchenstein
总规划：Gruner AG, Basel
视觉效果：Bloomimages, Hamburg
甲方：Kinderspital Zürich-Eleonorenstiftung, Zürich
总楼面面积：84,579m²
竣工时间：2018 (expected)

苏黎士儿童医院

Herzog & de Meuron

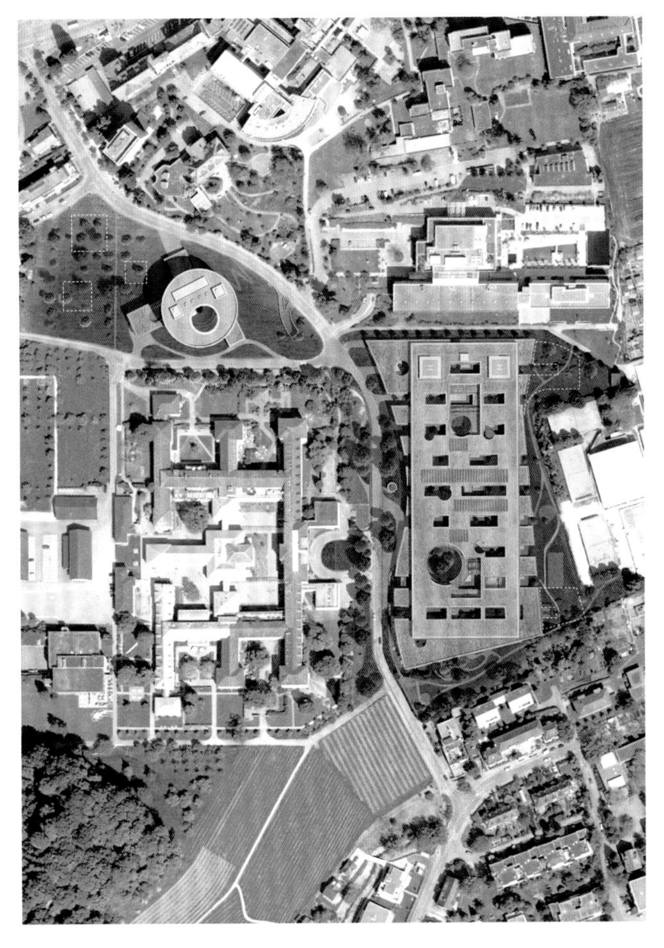

为了在苏黎士Lengg区建造一座新建筑，苏黎士Kinderspital委员会在2010年发起了一个需要进行资格预审的双阶段规划选拔程序。2012年3月，在评判委员会的推荐下，苏黎士Kinderspital基金委员会一致同意宣布赫尔佐格与德·梅隆建筑事务所的设计为获胜项目。该项目计划于2018年竣工。

儿童医院的设计想要建造两座互补的建筑，在建筑类型、规划以及城市设计方面形成对比，但在几何形状方面却有一定的关联性。医院主楼可为儿童及青少年提供检查和治疗服务，而教学与研究中心则可进行科学研究及调控工作。

儿童医院是一座环绕着一系列庭院排列的三层建筑，像一个小型的封闭城镇。住院病人、门诊病人及其家属可以在不同的治疗区域内自由穿行。这座三层建筑旨在建造一座适合儿童的医院大楼，在规模和材料方面可以让人联想到展馆式结构，并且不同于一般的医院模式。在这里，建筑与大自然相互交织在一起。内部庭院间歇性地向外开放，使光线可以照进低矮纵深的建筑中。

三层楼的每一层都有一个专属功能，这一点可以在房间的布局中体现出来，它们旨在为人们提供最大程度的灵活性。一层设有检查室、治疗室、实验室、治疗设施以及餐厅，二层设有医生办公室，而病房区则设在顶层。将这三层"理想的"几何结构叠加在一起就可以创造出空间的多样性与方向性。木材是立面和内部结构的主要材料，可以为孩子、家长及医护人员提供一个更加温馨的家庭氛围。对木材的使用也迎合了Lengg区周边的乡村环境。

教学与研究中心是一座高大的独立式圆形建筑，它位于康复园区中心的一块空地上，该园区是由几个不同的诊所组成的。显而易见，它是一座公共建筑，在这里所有的工作都是围绕科学研究及其知识的传播而展开的。研究实验室和办公室的六层建筑位于一个类似于广场结构的上方，该广场中设有会堂和研讨室。中央庭院是一个遵循太阳轨迹的圆形结构。

尽管这两种建筑类型截然不同，但它们在建筑结构上却有关联。它们二者的特点都是将矩形和圆形几何结构叠加起来。在儿童医院里，圆形结构可以用来避免直角的基本布局方式，并且可以标注出建筑的具体区域、方向点、分中心、会面点和过渡区域。圆形教学与研究中心本身就是一个会面点，在医院园区内形成了一个分中心。在其圆形的外壳中，房间都是呈直角的。

Zürich Children's Hospital

For the realization of a new building in the Lengg district of Zürich, Kinderspital Zürich called a two-phase planning selection procedure with pre-qualification in 2010. On May 3, 2012, at the recommendation of the jury, the foundation board of Kinderspital Zürich unanimously declared the design by Herzog & de Meuron as the winning project. The project is planned for completion in 2018.
The design for the children's hospital envisions two complementary buildings of contrasting typology, program and urban design that are nevertheless geometrically related. The main hospital building serves the examination and treatment of children and adolescents, while the teaching and research center serves scientific work and mediation.
The children's hospital takes the form of a 3-storey building arranged around a series of courtyards like a small, introspective town. In-patients, out-patients and their relatives can move around as freely as possible between the different treatment areas. The 3-storey reflects the desire to create a child-friendly building, reminiscent in both scale and materiality of pavilion-style structures, and differs the usual hospital stereotype. Architecture and nature are interwoven here. The interior courtyards open up intermittently to the outside, allowing daylight to permeate the

low and deep building.

Each of the three floors has a dedicated function, reflected in the layout of the rooms, and designed to provide maximum flexibility. Examination, treatment rooms and laboratories are located on the ground floor, as are the therapeutic facilities and the restaurant, while the doctors' offices are situated on the first floor, and the wards on the top floor. Overlaying these three "ideal" geometries creates spatial variety and orientation. Wood is the predominant material of the facades and interiors, creating a more domestic atmosphere for children, their parents and hospital staff. The use of wood also echoes the rural surroundings of the Lengg district.

The teaching and research center, a tall and round freestanding building, is positioned in an open space in the center of the health campus comprising several different clinics. It is an obviously public building in which everything revolves around scientific research and its dissemination. Six floors of research laboratories and offices are hovering above a kind of agora comprised of auditorium and seminar rooms. A central courtyard, also circular, follows the course of the sun.

Although the two building types are typologically different, they are architecturally related. Both feature an overlay of rectangular and circular geometries. In the children's hospital, the circle is used to interfere the right-angled basic order and to mark specific areas of the building, points of orientation, sub-centers, meeting points, transitional areas. The round teaching and research center is a meeting point in its own right, forming a sub-center on the hospital campus. Within its circular shell, the rooms are arranged at right angles. Herzog & de Meuron

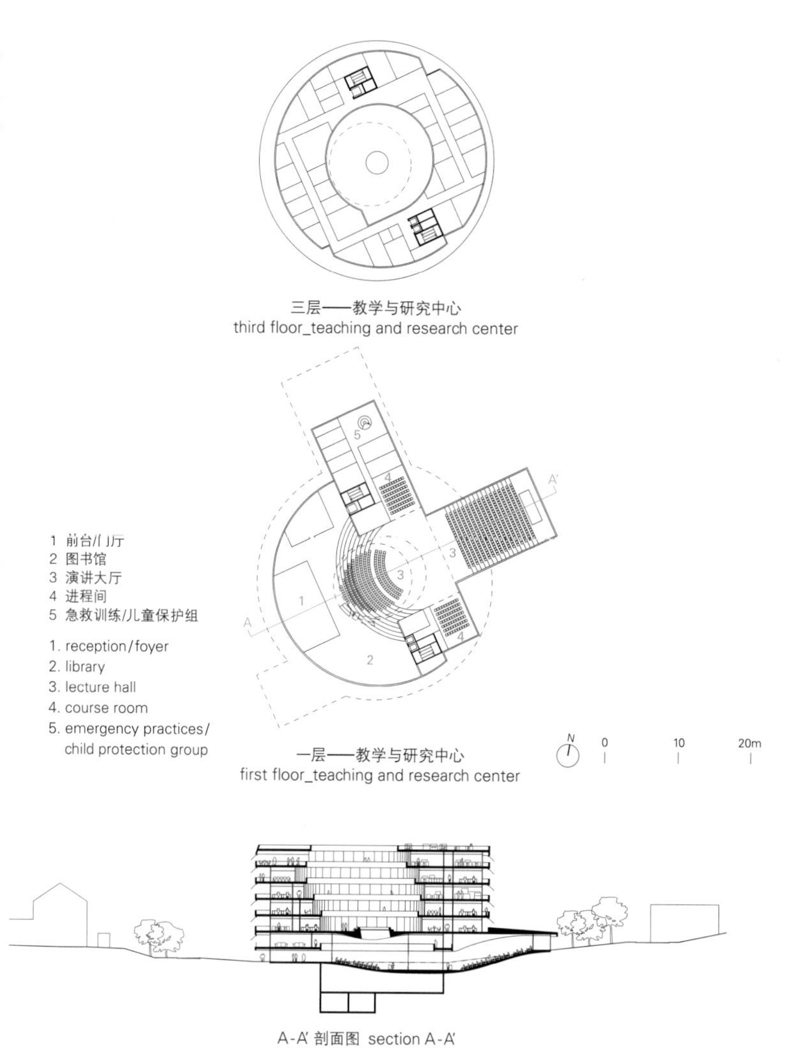

三层——教学与研究中心
third floor_teaching and research center

一层——教学与研究中心
first floor_teaching and research center

A-A' 剖面图 section A-A'

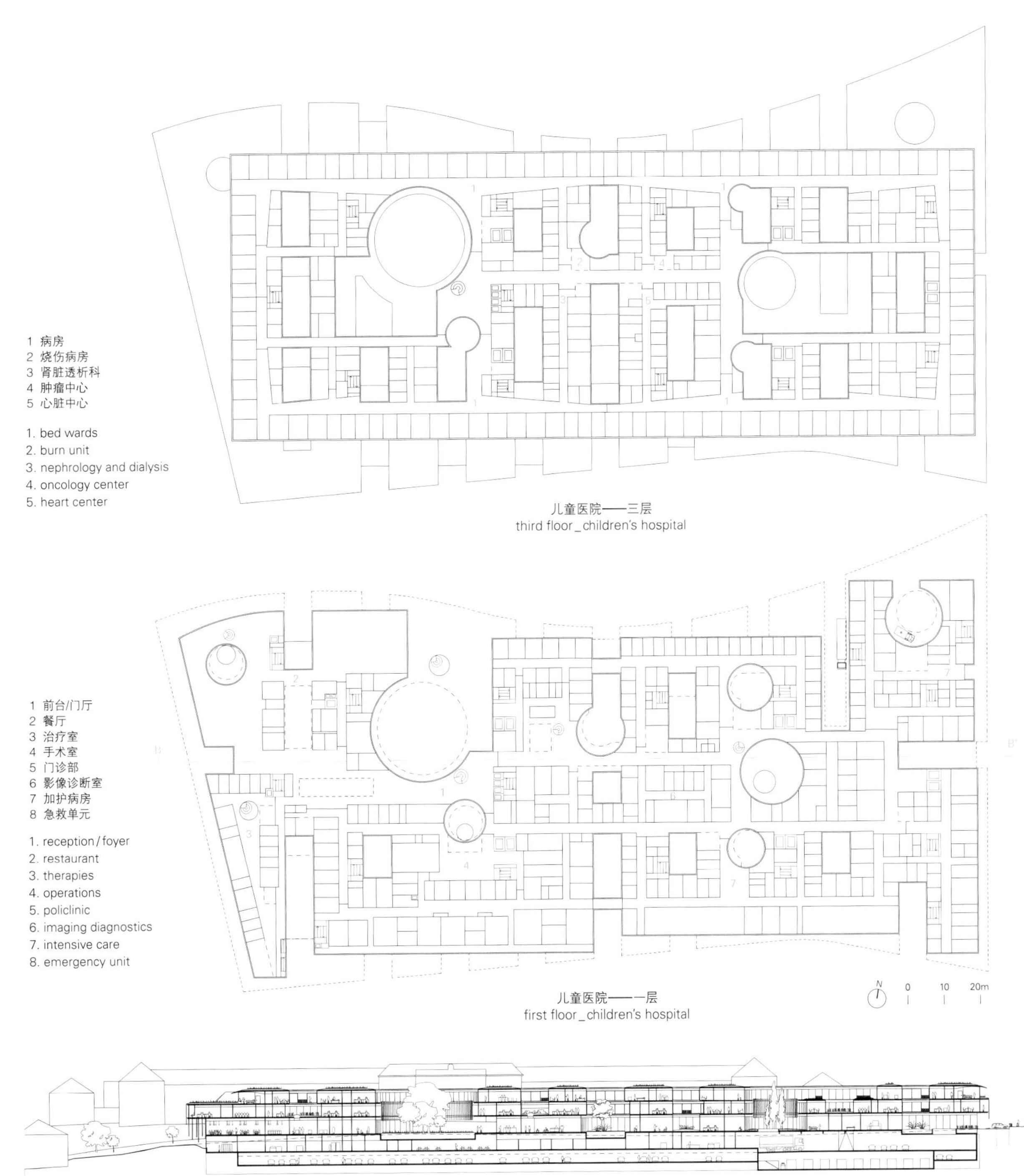

儿童医院——三层
third floor_children's hospital

儿童医院——一层
first floor_children's hospital

B-B′ 剖面图 section B-B′

项目：Santa Lucía University General Hospital
建筑师：C/ Mezquita, s/n, Paraje Los Arcos, 30202, Santa Lucía, Cartagena
开发商：Francesc Pernas, Bernat Gato, Roger Pernas
结构工程师：GISCARMSA, SAU
机械工程师：NB 35 SL, Jesús Jiménez, Fran Unzueta, Pedro Pastor
现场监理：Grupo J.G. SA, Joan Gallostra, Julio Morá
卫生与安全调度：Neoingena SL, José Mª Rueda, Iluminada Oliva, Francisco Aguilar
质量控制：SPGaudí
施工方：ACE Edificación
施工经理：UTE Hospital Cartagena(FCC + Intersa)
地点：Julio Alfaro
建筑面积：114.369m² 造价：EUR 121,360,653
设计时间：2003.5—2004.12 施工时间：2005.5—2010.7

圣卢西亚大学综合医院

Casa Sólo Arquitectos

为满足地中海沿海旅游区日益增长的需求，西班牙穆尔西亚自治区政府筹建了两座新的公立医院。该项目由一家特定的公司——The Murcia Health Service and Giscarmsa来经营，并且还举行了两次国际建筑竞赛。最终由Casa建筑事务所负责设计这两个项目：2005年的卡塔赫纳圣卢西亚医院和2006年的Los Arcos del Mar Menor医院。该项目从头到尾所有工作都由Casa建筑事务所负责，即：草图和详图设计、结构和工程项目的协调以及所有后续建设工程的监督问题，直至完成整个项目。

圣卢西亚医院外面有一层织物建筑包膜，使得其与周围的丘陵地貌协调。这层织物膜把建筑物的高度降到了更人性化的水平，还能使南立面防晒。并且它的颜色随着不同时间的阳光变化而变化，因此使得整个建筑物与环境很协调。

该医院还包括其他活动空间，如：运动、休闲和商业区。目的是使项目与社区更好地融合，以及进一步培养人们的健康观念。主要入口大厅没有暖气，提供了室外与室内之间的过渡，并且在此设置了若干娱乐消遣设施。

自然光透过网格结构中的庭院系统照在所有的设备上，在病人和工作人员有交集的区域阳光尤其充足。庭院的大小决定了它不仅可以用作天井，还可以给人们提供一个如同室外一样可以休息和会面的地方。护理区通过六个公共竖井与林荫大道直接连接，并与住院患者和医护人员的交通流线区隔开。访客和门诊病人可以乘坐电梯到二层的诊疗和门诊区以及三层绿色屋顶上的自助餐厅、运动、休闲及商业区域。

除了自然采光通风之外，该建筑的可持续性还通过一些其他设计特点来实现，比如：带厚厚的保温层和雨水收集功能的生态屋顶、将洗漱污水再利用来冲厕所、具有高保温性能的高效玻璃通风立面、脏衣物垃圾的气动传输以及屋顶的太阳能电池。

这家医院试图将医院和有利于维护健康的东西融合在一起，比如运动和休闲。还有一个设想就是对病房的创新性部署，以便于使病人得到更大的使用空间，同时还可以使医护人员通过走廊更清楚地看到病人的情况。Casa建筑事务所在参赛中提出的设计建议包括：健身房、跑道、泳池及商业区等等，所有这些都布局在中间的夹层，有些会布局在带有织物的创新遮蔽结构下面。这种结构也增强了该建筑物的可持续性，因为许多通往医院各种服务房间的夹层空间可以不用安装空调。

在医院低层和护理区之间夹有一层技术层，层高2.2m，这里装有所有空调机组和其他设备。该层可以通过一个交替天井系统获得穿堂风，可以吸收新鲜空气并排出废气。这样也更便于工程系统的维护以及未来的更新。

这种医院不仅是功能性建筑，同时还是备受关注的建筑作品。它使我们相信在建筑和设计中进行创新既对病人的健康有好处，还可以提高医护人员的工作效率。

Santa Lucia University General Hospital

The government of the Murcia region in Spain has developed two new public hospitals to fulfill the increasing demand of its coastal touristic areas in the Mediterranean Sea. The Murcia Health Service and Giscarmsa, a specific company created to run the projects, launched two international architecture competitions. Casa was awarded these two projects: Santa Lucia Hospital in Cartagena in 2005 and Los Arcos del Mar Menor Hospital in 2006. Casa had been responsible for the full process: sketch and detailed design, structural and engineering project coordination, and follow-up control of all building stages until works have been completed.
The Santa Lucia hospital has a textile architectural envelope, shaped to get along with the hilly surrounding landscape. This faceted screen cuts the building height to a more human scale and gives solar protection to the southern facades. Its color's changes with daylight at different times achieve the total integration of such a big structure with the environment.
The program includes other activities, such as sports, leisure and commercial areas. Its objective is to get a stronger integration of the community into the hospital building area, and to develop a better sense of health prevention. The general access concourse

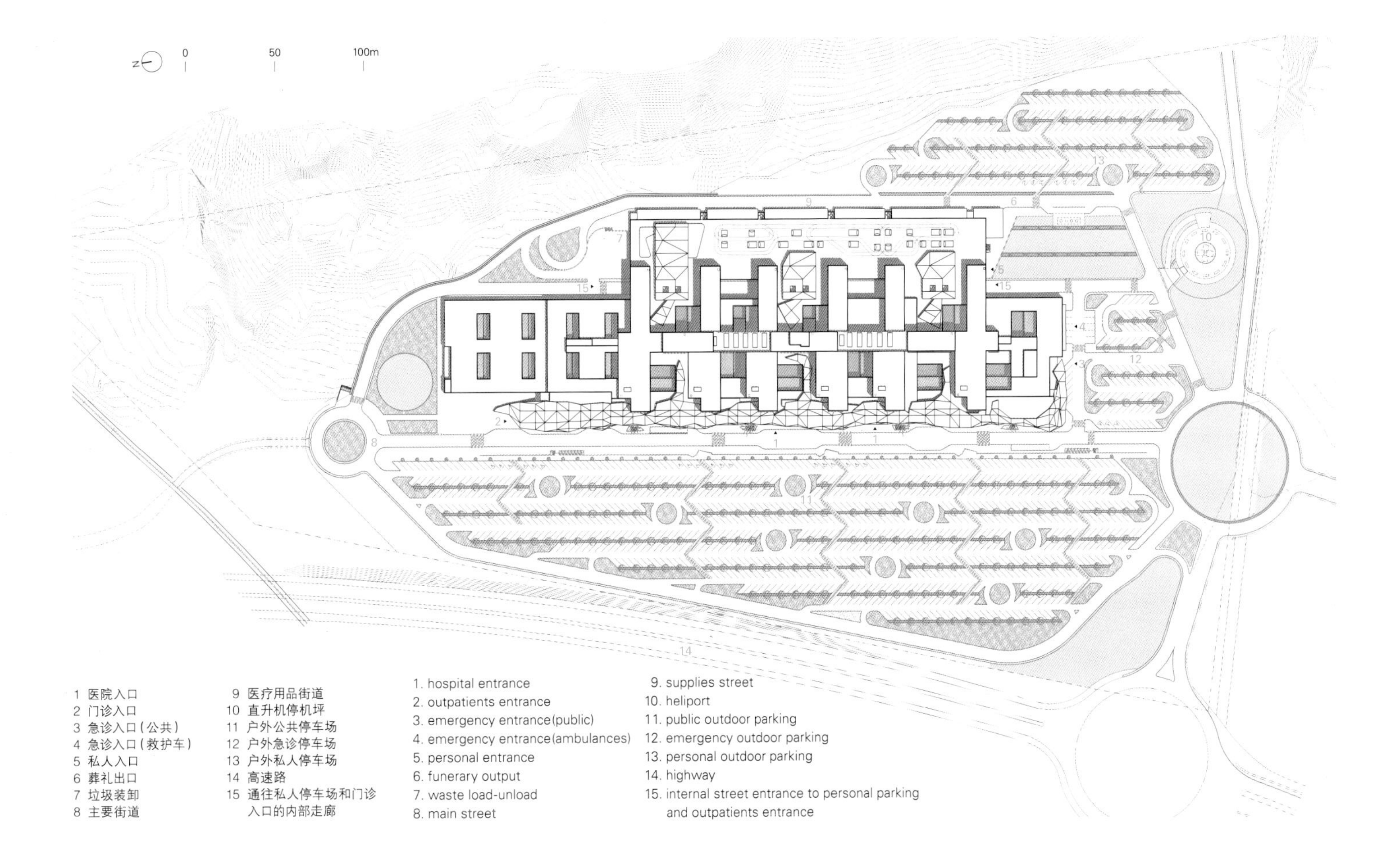

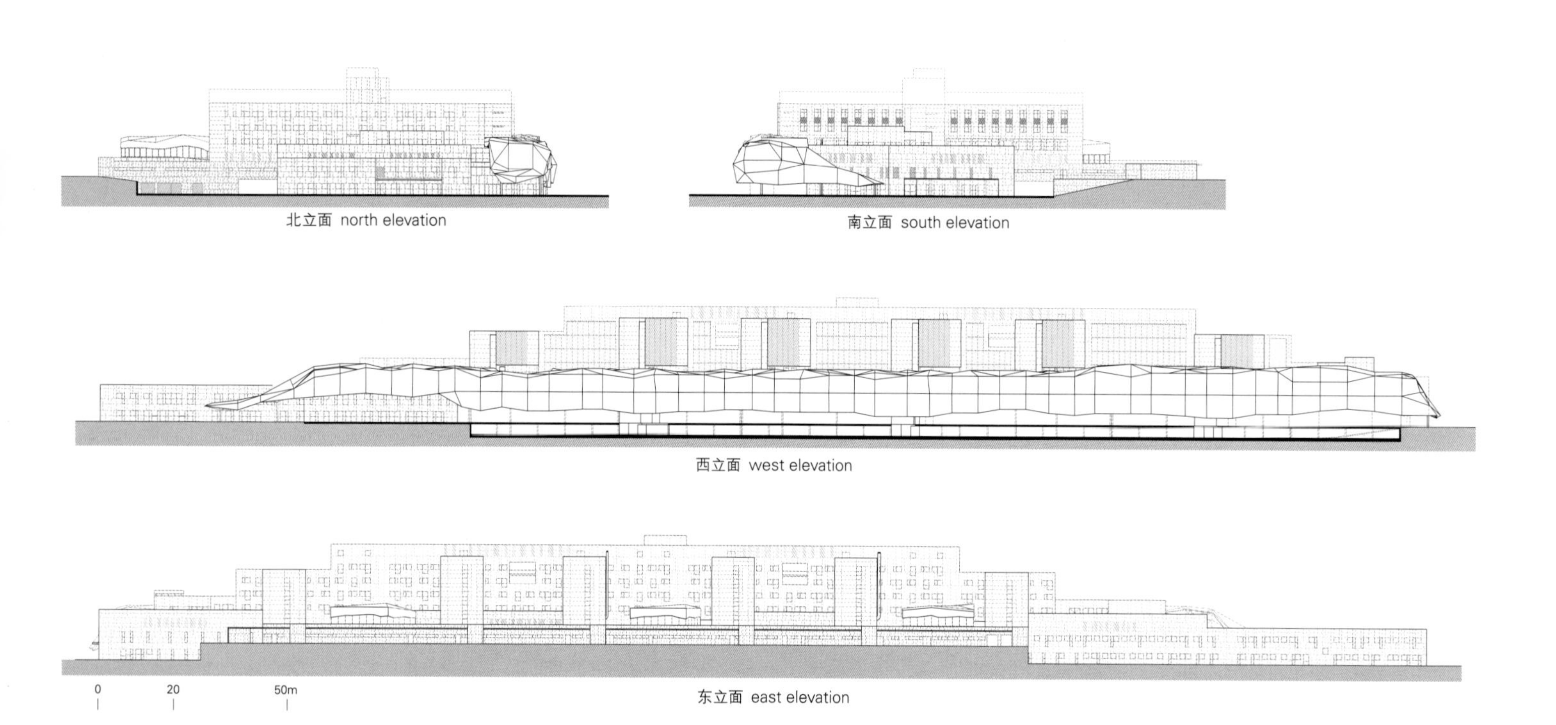
北立面 north elevation
南立面 south elevation
西立面 west elevation
0
20
50m
东立面 east elevation

Hospital General Universitario

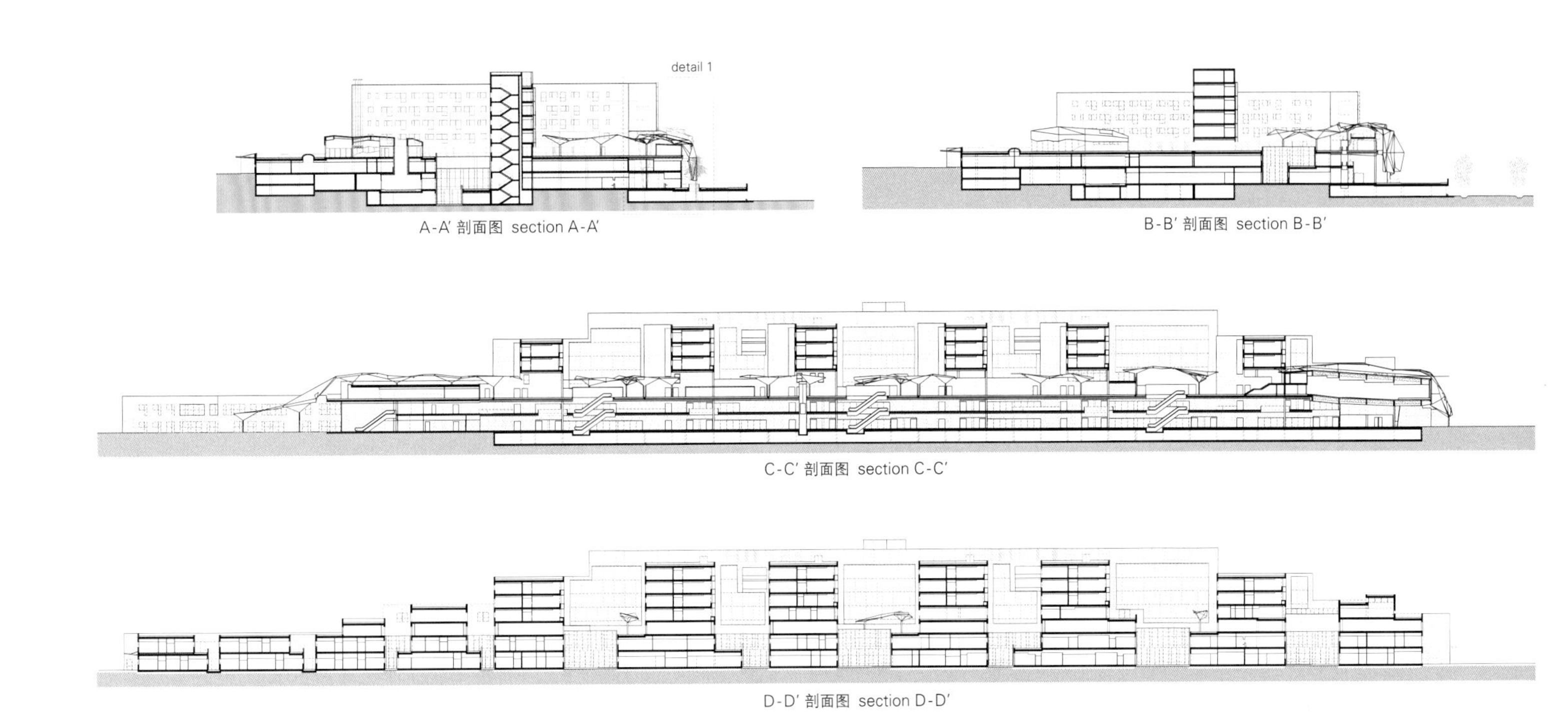

A-A' 剖面图 section A-A'

B-B' 剖面图 section B-B'

C-C' 剖面图 section C-C'

D-D' 剖面图 section D-D'

is an unheated umbraculum providing a transition between exterior and interior and is also where leisure facilities are located.
Natural light reaches all services through a courtyard system perfectly fitted into the structural grid, particularly in those spaces of relationship between patients and staff. The size of these courtyards allows them to be used not only as light wells, but also as staying external areas for relax and meeting. Nursing units are directly connected with the boulevard by six public vertical shafts, separated from inpatients and staff circulations. The elevators bring also visitors and outpatients to the diagnostic and treatment units situated on the first floor, and to the cafeteria, sports, leisure, and commercial spaces situated on the second floor's green roof.
Besides natural lighting and ventilation, the sustainability of the buildings is achieved with several other design features such as: ecological roofs including thick insulations and rainwater collection, recuperation and of some grey water to be used in WC's flushing, ventilated facades with high insulation performances and high efficiency glazing, pneumatic transportation of dirty linen and trash, solar cells either on the roof.
This hospital is an attempt to incorporate into the hospital, some other programs with a certain relationship with health maintenance, such as sport and leisure. There was also the idea to explore a new disposition of the patient's room, both to gain more space for patient's use and to allow better patient's view from corridor by nursing staff. The proposal submitted by Casa to the competition included a fitness center, a running path, swimming pool, commercial areas, all of them on an intermediate roof, and some under a very new concept of the shadowing structure incorporating textiles. This structure is also reinforcing the sustainability of the building, because many of the intermediate spaces of access to the different hospital services can be designed without air-conditioning.
Between the lower levels and the nursing units, there is an intermediate technical floor of 2.2 meters height including all HVAC units and other plants. They receive cross ventilation through an alternate patio system to get fresh air and discard it after use. It facilitates maintenance and future changes of the engineering systems.
Such hospitals are to be considered not only as functional buildings, but also as interesting architectural works. It makes us to believe that innovation in architecture and design contributes also to patient's wellness, and staff attraction to work in those hospitals. Casa Sólo Arquitectos

1. estructura depórticos de hormigón armado
2. forjado reticular con casetón de hormigón no recuperable
3. muros de conteción encofrados a doble cara
4. forjado de losa maciza de hormigón armado
5. estructura metálica formado por perfiles tubulares huecos
6. estructura metálica formada por perfiles de acero laminado tipo IPE
7. forjado dechapa colaborante de acero galvanizado con capa de compresión de hormigón armado
8. aplacado piedra arenisca de 4cm de grueso en grandes formatos sobre estructura de acero galvanizado
9. protector solar textil microperforado tipo Precontraint 371 o equibalente
10. muro cortina T6. Paño superior. vidrio traslucido
11. pared de bloque de mortero de cemeto
12. pavimento PVC tarket o equivalente sobre capa de mejora / A1*. conductivo
13. pintura epoxi antipolvo
14. pavimento gres Ston-ker o equivalente
15. aplacado tablero de fórmica a altura falso techo con zocalo inox
16. pintado al plástico liso
17. techo cartón-yeso liso no registrable
18. recrecido de mortero
19. capa arena - cemento 4cm
20. barandilla de perfiles de acero inox y vidrio laminado
21. plancha de acero lacado
22. capa niveladora
23. bordillo hormigón prefabricado 100x20x20cm
24. rigola 20x20x8cm
25. subbase de grava
26. lámina impermeable de polietileno
27. solera de hormigón armado
28. escalera mecánica de 80cm de paso y 35° inclinación
29. acabado de grava; hormigón, celula de espesor medio 10cm(solo en caso de pendiente 1%), capa de mortero fratasado 2~3cm, geotextil 300, lámina impermeable FPO1.5mm, poliestireno extrusionado 40mm, capa separadora polipropileno 400, acabado de grava
30. cubierta flontante sobre plots acabado madra; mortero de formación de pendientes 2~3cm, geotextil 300, lámina impermeable FPO 1.5mm, poliestireno extrusionado 40mm, capa separadora de polipropileno 400, soporte SRE / U rased o equivalente para tarimas de madera, tarima de madera de pino tratado con junto abierta 10mm de 190x40mm clavado a rastrel
31. cubierta flontante sobre plots acabado gres Ston - ker o equivalente; mortero de formación de pendientes 2~3cm, geotextil 300, lamina impermeable FPO 1.5mm, poliestireno extrusionado 40mm, capa separadora de polipropileno 400, soporte SRE/U rased o equivalente para tarimas de madera, acabado gres Ston-ker o equivalente
32. cubierta calle calzada; mortero acabado fratasado 8cm, geotetil 300, lamina impermeable FPO 1.5mm, capa de protección de altas prestaciones 8mm, capa separadora de polipropileno 400, protección de hormigón 5cm, aglomerado asfáltico
33. cubierta calle acera; mortero acabado fratasado 8cm, geotetil 300, lamina impermeable FPO 1.5mm, capa de protección de altas prestaciones 8mm, capa separadora de polipropileno 400, grava, capa de mortero, pavimeto exterior de gres Ston-ker o equivalente
34. remate de antepecho de plancha de acero lacado 0.7mm
35. pavimento de madera para exterior con junta abierta
36. tabique de carton-yeso

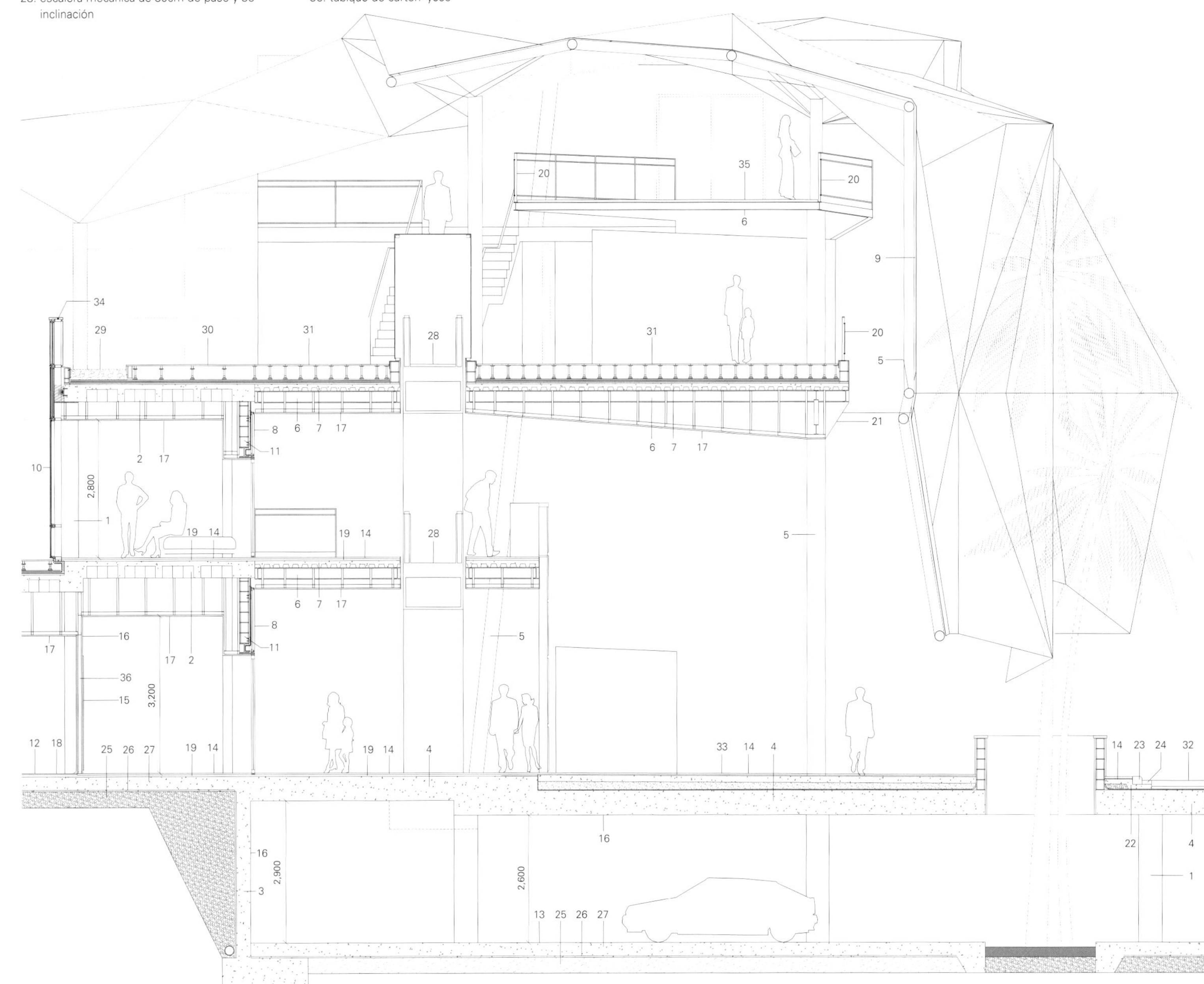

详图1 detail 1

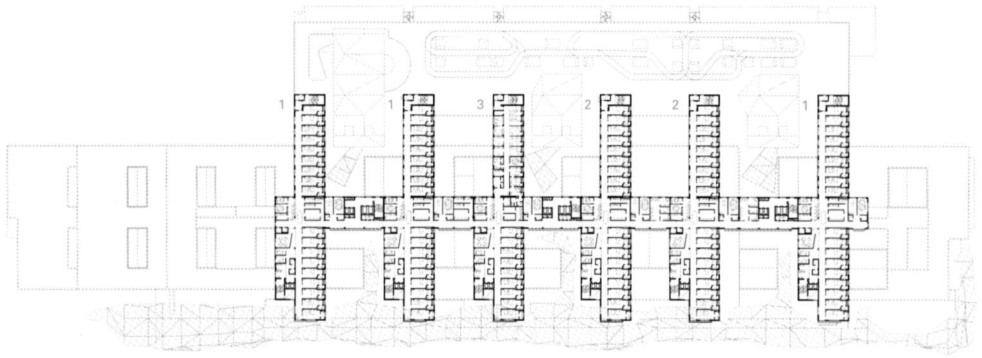

四层 fourth floor

1 普通住院处
2 产科住院处
3 产科住院处/新生儿重症监护室

1. conventional hospitalization unit
2. obstetrics hospitalization unit
3. obstetrics hospitalization unit/ neonatology-NICU

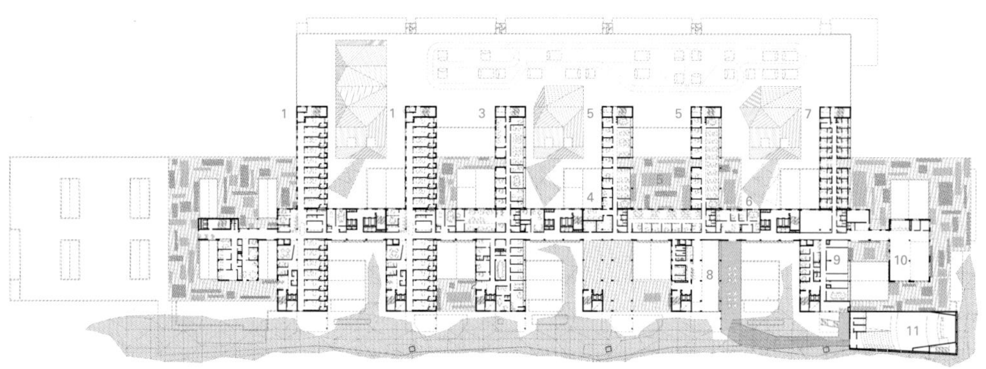

三层 third floor

1 儿科住院处
2 短期住院处
3 指挥管理办公室
4 信息处
5 行政管理空间
6 健康研究室
7 警卫室
8 员工餐厅
9 教室
10 图书馆
11 礼堂

1. pediatric hospitalization unit
2. short-term hospitalization
3. direction and administration
4. informatics
5. administrative space
6. health laboratory
7. personal residence of the guard
8. cafeteria for staff
9. teaching unit
10. library
11. assembly hall

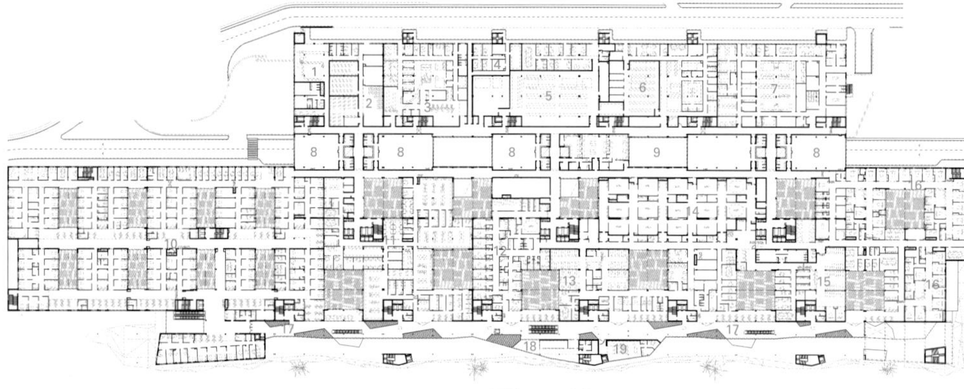

二层 second floor

1 保洁处
2 被服库
3 厨房
4 信息处
5 医疗用品储藏室
6 药房
7 病理解剖室
8 员工更衣室
9 灭菌室
10 门诊部
11 实验室
12 产科
13 日间医院手术室
14 急救手术与医疗手术室
15 麻醉复苏室
16 特护病房
17 林荫道
18 娱乐区
19 礼拜堂

1. cleaning
2. linen shop
3. kitchen
4. informatics
5. supplies and storage
6. pharmacy
7. pathological anatomy
8. staff changing room
9. sterilization
10. outpatient
11. laboratories
12. obstetrical block
13. surgical day hospital
14. ambulatory surgical and medical surgery block
15. resuscitation post-anesthetic
16. intensive care unit
17. boulevard
18. leisure
19. church

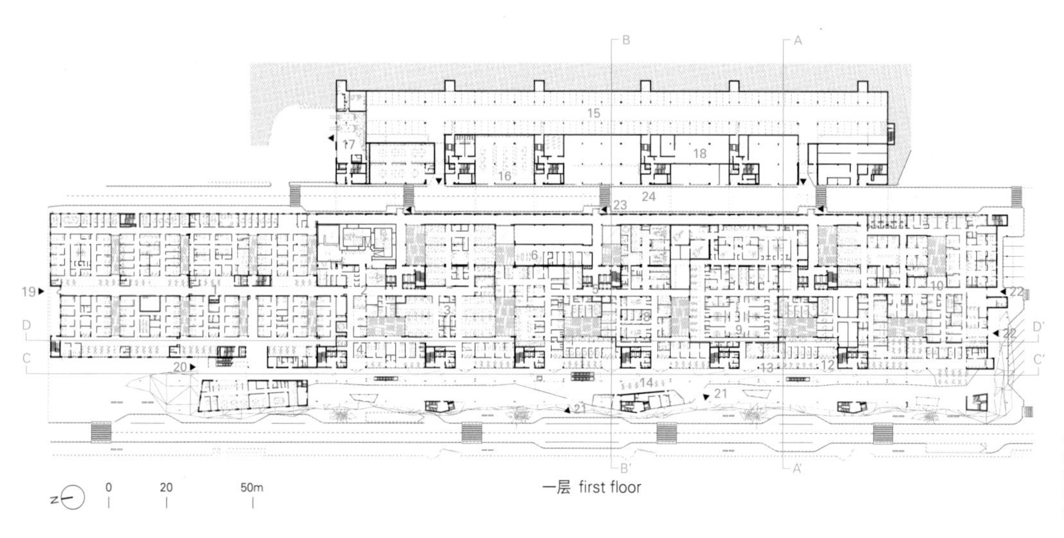

一层 first floor

1 门诊部
2 放疗室
3 日间医院医疗室
4 疼痛观察室
5 透析室
6 复健治疗室
7 血液动力室
8 核医学室
9 影像室
10 急诊室
11 住院部
12 护工服务与社会工作处
13 药房
14 林荫道
15 员工停车场
16 维修处
17 垃圾站
18 设备处
19 限制使用通道
20 外部咨询入口
21 医院入口
22 急诊入口
23 门诊区内部通道
24 门诊区内部走廊

1. outpatient
2. radiotherapy
3. medical day hospital
4. pain unit
5. dialysis unit
6. rehabilitation
7. hemodynamic
8. nuclear medicine
9. imaging
10. emergency
11. admissions
12. patient care services and social work
13. pharmacy
14. boulevard
15. staff parking
16. maintenance
17. waste
18. facilities
19. restricted access
20. access external consultations
21. hospital access
22. emergency access
23. internal access outpatient areas
24. internal street outpatient areas

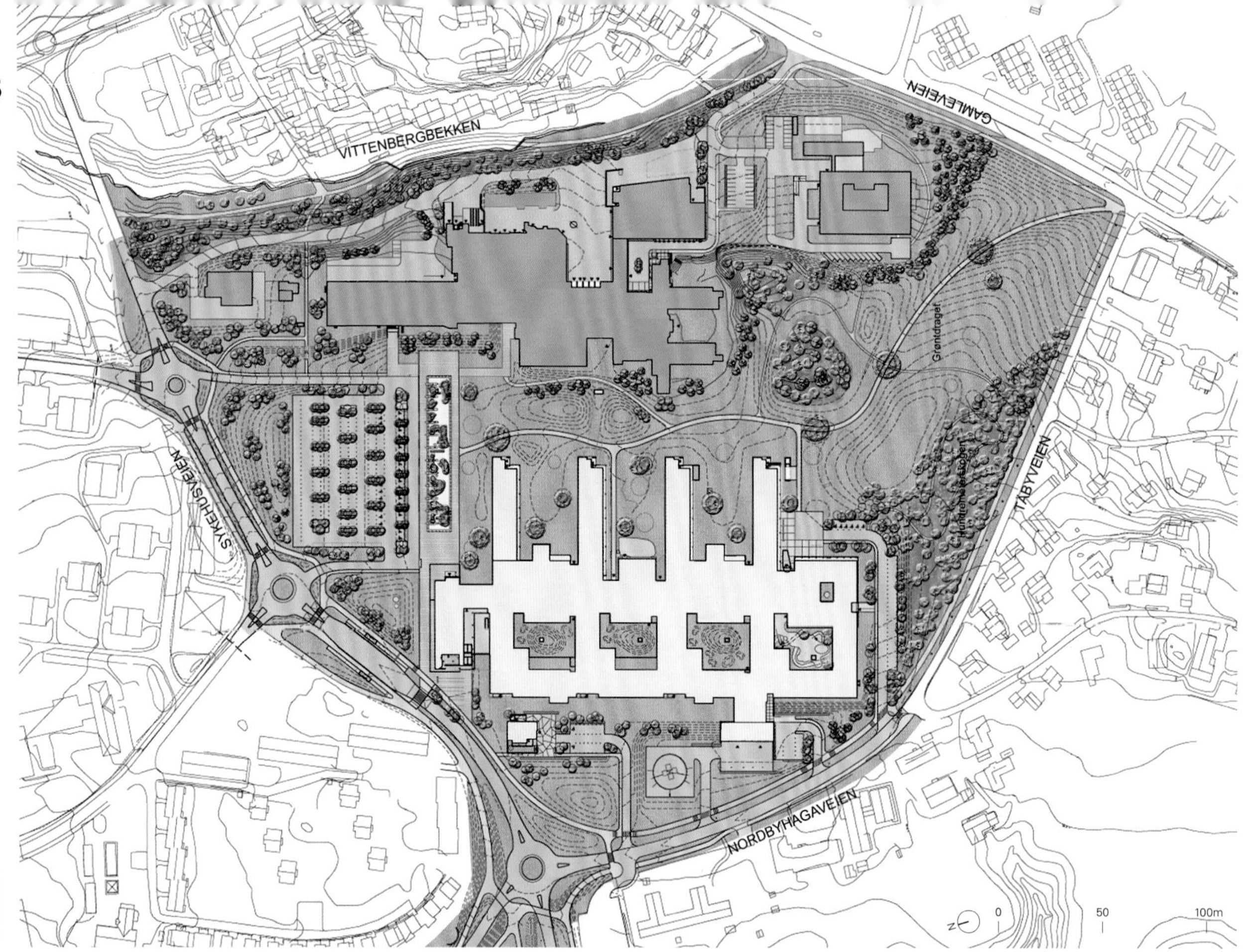

阿克什胡斯大学医院

C.F. Møller Architects

阿克什胡斯大学医院紧邻奥斯陆，它不是一所传统的机构建筑，而是一个友好、闲适的场所，周围呈开放式且井井有条。这种周边环境使患者及其家属有种亲切感。阿克什胡斯大学医院的设计意在强调其在周边丰富的环境中的安全性和清晰性，同时周边的日常功能和熟知的材料也被融合进医院的结构中。

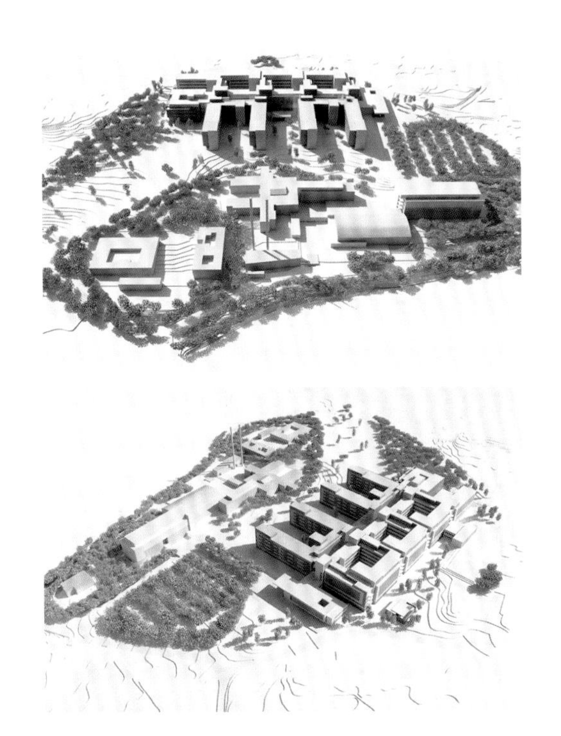

统一性和变化

建造过程中所用的材料变化多样。然而这种多样性在以面板和透明性为中心的总体建筑主题下变得统一。这样，建筑综合体的各个部分就被整合成一个整体，也由此产生一种透明而深刻的微妙效果。中央以玻璃覆顶的主干道将不同的建筑和科室联系起来。这条玻璃通道从到达处（设有主接待台接待来访者）亲切的大厅开始，延伸了几百米长，最后回到大厅和儿童科的独立到达区。

像小镇一样布局

玻璃通道是该建筑具有统一作用的构件。这里，多种当地材料——木材、石头和玻璃——都统一在建筑的各处。由冰岛艺术家Birgir Andrésson设计的大型彩色面板成为一种自然元素，并为整座医院的配色方案提供了画板。艺术在阿克什胡斯大学医院随处可见。它以多种形式、外观、规模被融入建筑，并有意与传统医院建筑的功能性形象形成对比。玻璃通道为建筑提供了一个像小镇一样的结构，带有小型广场和开放式空间。在玻璃通道内部人们可以看到小镇的日常功能设施：教堂、药房、美发店、花铺、咖啡厅、电话亭、交通节点以及其他服务性设施，以方便病人、家属和医院员工使用。

以病人为中心

所有的治疗病房都沿着玻璃通道的一侧排布，以四个庭院为中心。这些布局保证病人可以拥有良好日常生活，同时还可有效管理病人与外界、员工的接触。所有病房都沿着玻璃通道的西南侧布置，既可以拥有充足的自然光照也能看到景观中的美景。儿童科的病房上装配的窗户使儿童和青少年在他们的病床上就能看到天空和周围植被的独特景色。为家长准备的设备精良的装置保证儿童和其家庭沟通无碍。阿克什胡斯大学医院结构设计中最重要的特征是整个建筑综合体对高质量自然光的集中需求，该需求通过功能丰富的主干道的玻璃屋顶、通道中部的大型层压玻璃构件以及病房和治疗科的大型窗户来满足。该特征与医院周边环境——庭院中苔绿色、长满苔藓的平台，当地的花岗岩结构和附近的田野及树林景观——产生了强烈互动。

可持续性理念

阿克什胡斯大学医院拥有高度可持续性的设计，采用了当地材料和地热能源，满足了医院85%的供暖需求和超过40%的总能耗。各功能设施之间的距离很近，医院布局也很清晰，同时还广泛采用了包括机器人在内的现代科技，这些都使员工可以把更多时间花在病人身上。

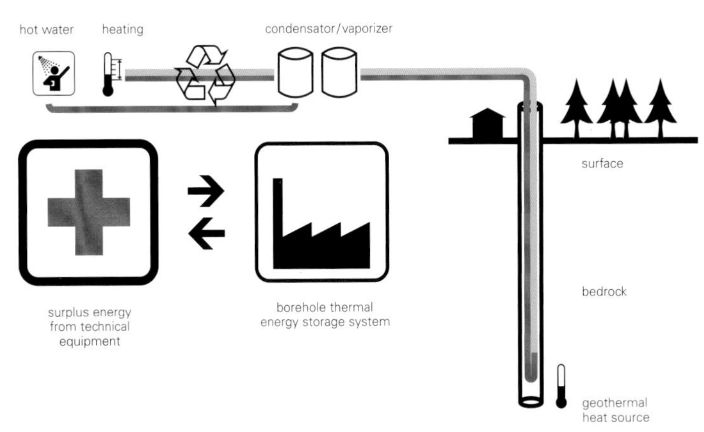

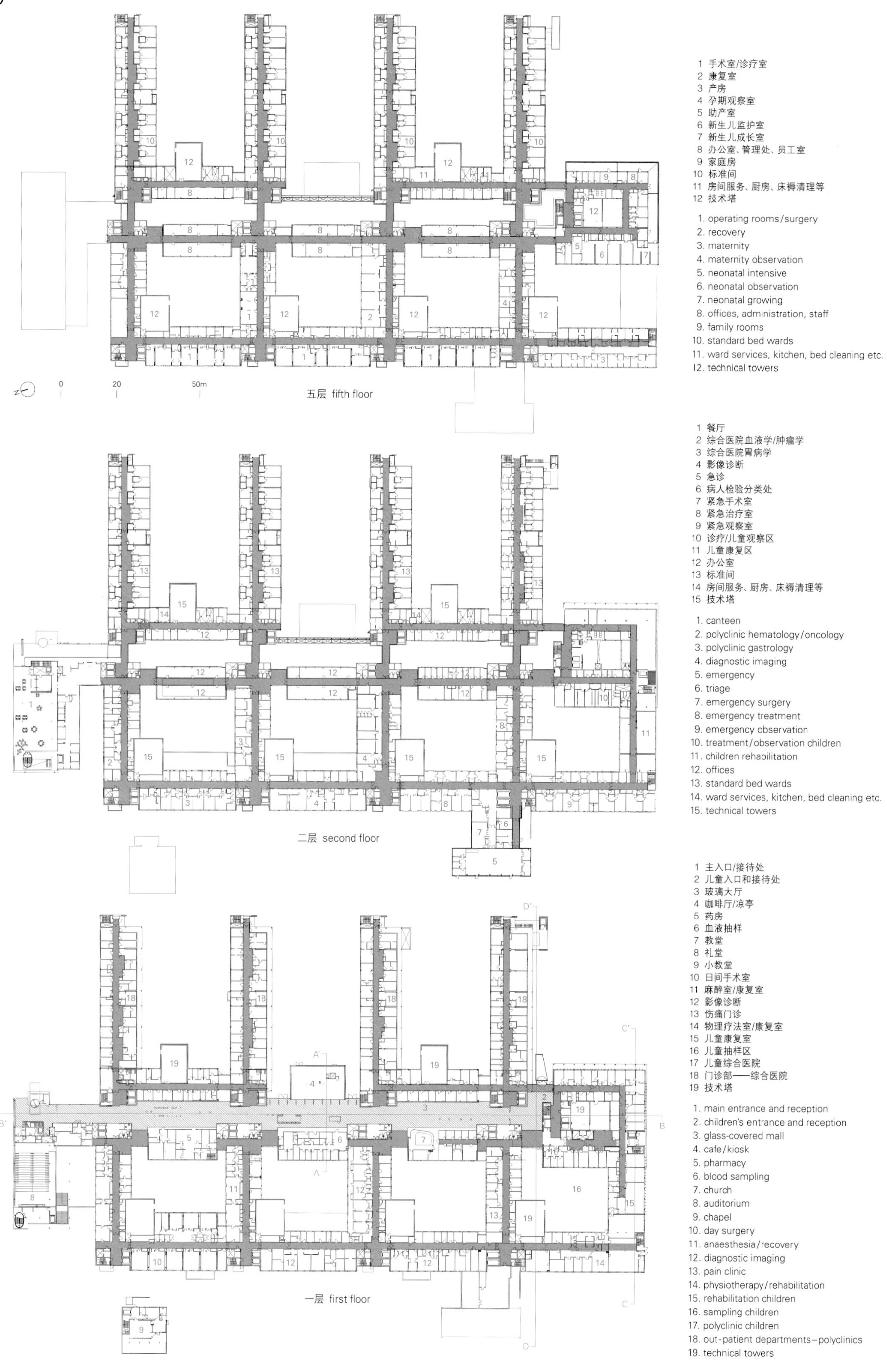
1 手术室/诊疗室
2 康复室
3 产房
4 孕期观察室
5 助产室
6 新生儿监护室
7 新生儿成长室
8 办公室、管理处、员工室
9 家庭房
10 标准间
11 房间服务、厨房、床褥清理等
12 技术塔
1. operating rooms/surgery
2. recovery
3. maternity
4. maternity observation
5. neonatal intensive
6. neonatal observation
7. neonatal growing
8. offices, administration, staff
9. family rooms
10. standard bed wards
11. ward services, kitchen, bed cleaning etc.
12. technical towers
0
20
50m
五层 fifth floor
1 餐厅
2 综合医院血液学/肿瘤学
3 综合医院胃病学
4 影像诊断
5 急诊
6 病人检验分类处
7 紧急手术室
8 紧急治疗室
9 紧急观察室
10 诊疗/儿童观察区
11 儿童康复区
12 办公室
13 标准间
14 房间服务、厨房、床褥清理等
15 技术塔
1. canteen
2. polyclinic hematology/oncology
3. polyclinic gastrology
4. diagnostic imaging
5. emergency
6. triage
7. emergency surgery
8. emergency treatment
9. emergency observation
10. treatment/observation children
11. children rehabilitation
12. offices
13. standard bed wards
14. ward services, kitchen, bed cleaning etc.
15. technical towers
二层 second floor
1 主入口/接待处
2 儿童入口和接待处
3 玻璃大厅
4 咖啡厅/凉亭
5 药房
6 血液抽样
7 教堂
8 礼堂
9 小教堂
10 日间手术室
11 麻醉室/康复室
12 影像诊断
13 伤痛门诊
14 物理疗法室/康复室
15 儿童康复室
16 儿童抽样区
17 儿童综合医院
18 门诊部——综合医院
19 技术塔
1. main entrance and reception
2. children's entrance and reception
3. glass-covered mall
4. cafe/kiosk
5. pharmacy
6. blood sampling
7. church
8. auditorium
9. chapel
10. day surgery
11. anaesthesia/recovery
12. diagnostic imaging
13. pain clinic
14. physiotherapy/rehabilitation
15. rehabilitation children
16. sampling children
17. polyclinic children
18. out-patient departments–polyclinics
19. technical towers
一层 first floor

项目名称：Akershus University Hospital
地点：Oslo, Norway
建筑师：C. F. Møller Architects
合作者：Multiconsult AS, SWECO AS, Hjellnes COWI AS / Interconsult ASA, Ingemannson Technology, Nosyko/Erstad og Lekven
景观建筑师：Bjorbekk & Lindheim AS, Schonherr Landskab A/S
艺术指导：Troels Wörsel, Gunilla Klingberg, Mari Slaattelid, Knut Henrik Henriksen, Jan Christensen, Tony Cragg, Birgir Andrésson, Petteri Nisunen, Tommi Grönlund, Julie Nord, Per Sundberg, Vesa Honkonen, Janna Thöle-Juul, Kristine Halmrast, Mikkel Rasmussen Hofplass
甲方：Helse Sør-Øst RHF
总楼面面积：137,000m² (118,000m² new build)
施工时间：2000—2008
摄影师：©Torben Eskerod (courtesy of the architect)-p.48~49, p.52, p.53[left], p.54, p.56, p.57
©Guri Dahl (courtesy of the architect)-p.51, p.53[right]

可再生能源

医院每年的总能耗接近20GWh，与1300家独户住宅一年消耗的能量相仿。可再生能源的使用基于一个地面热交换系统和基岩中的热储存能力。基岩中的多余热量（例如来自太阳得热、人体、技术设备、制冷和通风设备的热量）可以储存于350口200m深的能量井中。

Akershus University Hospital

The Akershus University Hospital just outside Oslo is not a traditional institutional construction; it is a friendly, informal place with open and well-structured surroundings. These surroundings present a welcoming aspect to patients and their families. Akershus University Hospital has been designed to emphasize security and clarity in its rich surroundings, where everyday functions and well-known materials are integrated into the hospital's structure.

Wholeness and Variation

The material expression of the development is rich in variation. Nonetheless this expression is united into a whole by means of a general architectural theme centred on panels and transparency. In this way, a unity is created between the individual parts of the complex, which thereby receive a subtle effect of transparency and depth. A central, glass-roofed main thoroughfare links the various buildings and departments. This glass street begins in the welcoming foyer of the arrivals area, where the main reception desk receives visitors. The street runs several hundred metres in length and concludes in the foyer and the separate arrivals area of the children's department.

Structured like a Town

The glass street is the unifying part of the development. Here, various local materials – wood, stone, and glass – are united in an overall composition. Large colored panels designed by Icelan-

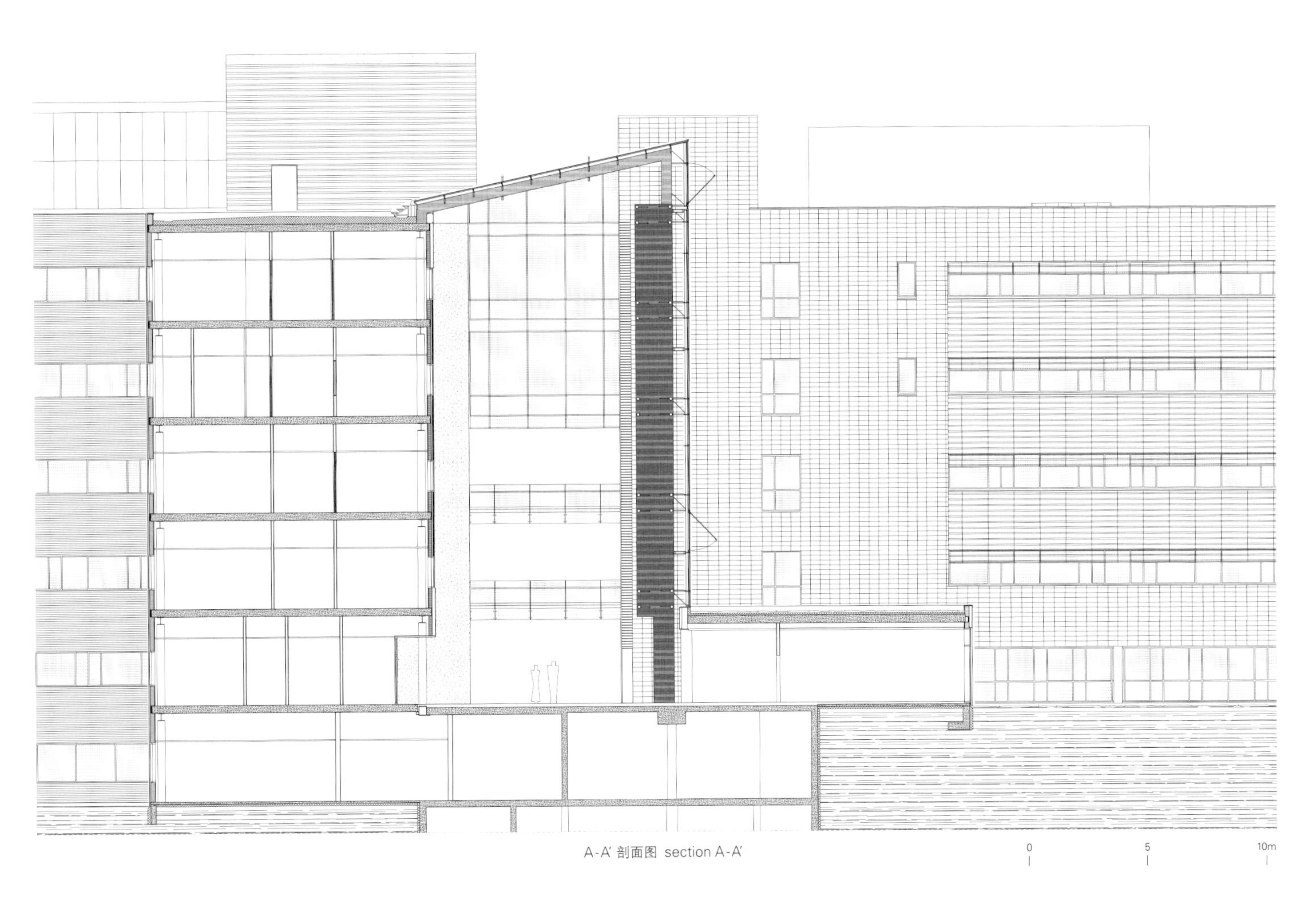

A-A' 剖面图 section A-A'

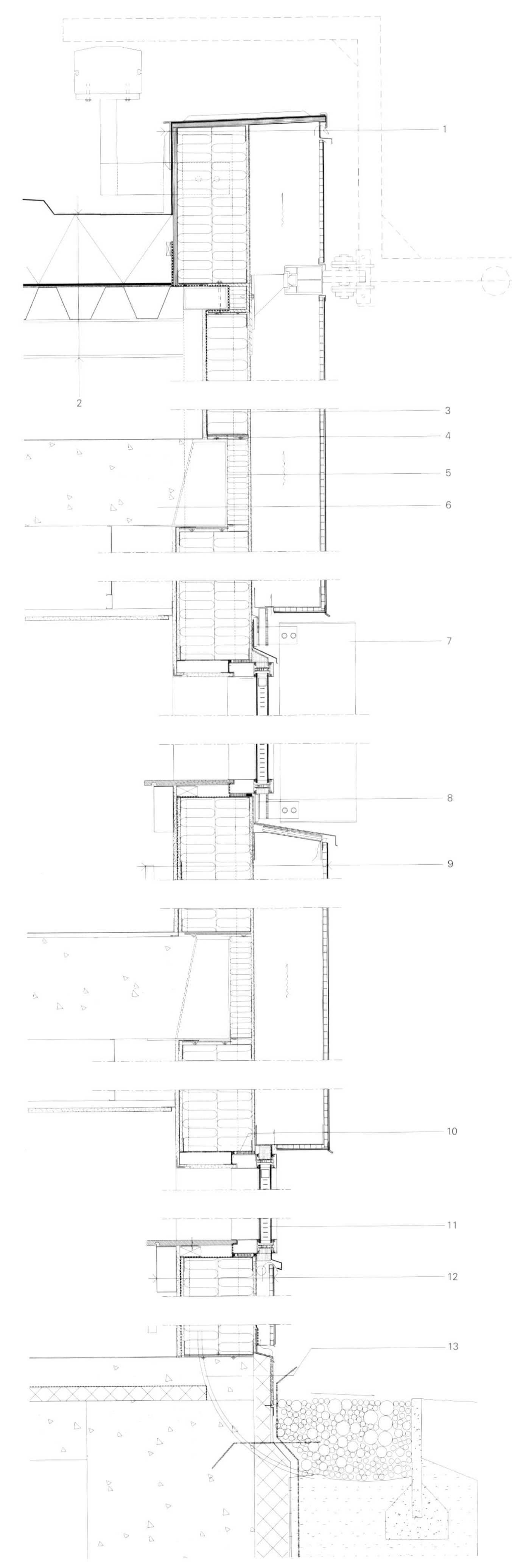

诊疗大楼详图 treatment building detail

dic artist Birgir Andrésson form a natural element and provide a palette for the color scheme of the whole hospital. Art is visible everywhere throughout the Akershus University Hospital. It has been integrated in many forms, shapes, and sizes and creates a deliberate contrast to the traditional, functional image of hospital construction. The glass street provides a town-like structure with squares and open spaces. In the glass street you'll find the everyday functions of such a town: church, pharmacy, hairdressers, florists, cafe, and kiosk, as well as traffic nodes and other services for the benefit of patients, relatives, and staff.

Centered on the Patient

All treatment wards are located on one side of the glass street and centred on four courtyards. These ensure a well-defined daily life for patients, with a manageable level of social contact and contact to staff. All bed wards are located to the southwest side of the glass street providing them with ample daylighting levels and beautiful views to the landscape. The wards of the children's department are equipped with windows which give the children and young people individual views of both the sky and the surrounding greenery from their beds. The well-equipped facilities for parents secure excellent contact between the children and their families. The most important characteristic in the physical design of the Akershus University Hospital has been the central requirement for high quality daylighting throughout the entire complex – right from the rich main thoroughfare, via glass roofs and the impressive glue-lam and glass sections in the middle of the street, to the generously-sized windows of the wards and treatment departments. This creates a strong interplay with the hospital's surroundings, from the mossy green, lichen-covered terrain in the courtyards to the views of the local granite formations and nearby fields and woodlands.

Sustainability

The Akershus University Hospital is a highly sustainable design, making use of locally sourced materials, and geo-thermal energy to provide 85% of the hospital's heating and more than 40% of the total energy consumption. Short distance between functions, a clear organisation and extensive use of modern technology including robotics give staff more time for patients.

Renewable Energy Sources

The total energy consumption of the hospital is approx. 20 GWh/year, similar to the consumption of 1300 single family houses. The use of renewable energy is based on a ground heat exchange system, combined with thermal storage capacity in the bedrock, where surplus heat (for instance from solar gain, people, technical equipment, cooling and ventilation plants) can be stored in 350 energy wells drilled to a depth of 200m.

C.F. Møller Architects

1. window cleaning system
   standing seam flashing, slope 1:20
   15x135mm opening for air circulation
   2mm perforated aluminium flashing
   with netting; rubber backing rod
   parapet attached to column
   HUP 180x180
2. parapet attached to column
   steel prepared for curtain wall
   bituminous roofing membrane
   250~400mm load bearing thermal insulation
   load distributing steel plate
   plastic vapor barrier 0.2mm; TRP 111 deck
3. HUP 180x180 column
4. cell rubber backing rod
5. steel beam
6. poured-in-place concrete slab
7. vertical fixed louver of extruded
   aluminium bxd 50x270mm
   end plates welded to top and bottom
8. cement fibreboard attached to metal profile
9. stucco on 12mm mineral fibreboard
   250mm vertical steel furring c/c 600mm
   corrosion classification C3
   weather barrier
   9mm exterior sheathing
   2x125mm thermally insulated framing
   metal band for radiator attachment
   13mm gypsum; vapor barrier
10. elastic sealant, backing rod and mineral wool
11. window with built-in venetian blinds
12. stucco on 12mm mineral fibreboard
    50 mm vertical steel furring c/c 600mm
    corrosion classification C3
    weather barrier
    9mm exterior sheathing
    2x125mm thermally insulated framing
    metal band for radiator attachment
    13mm gypsum; vapor barrier
13. conduit for lightning rod

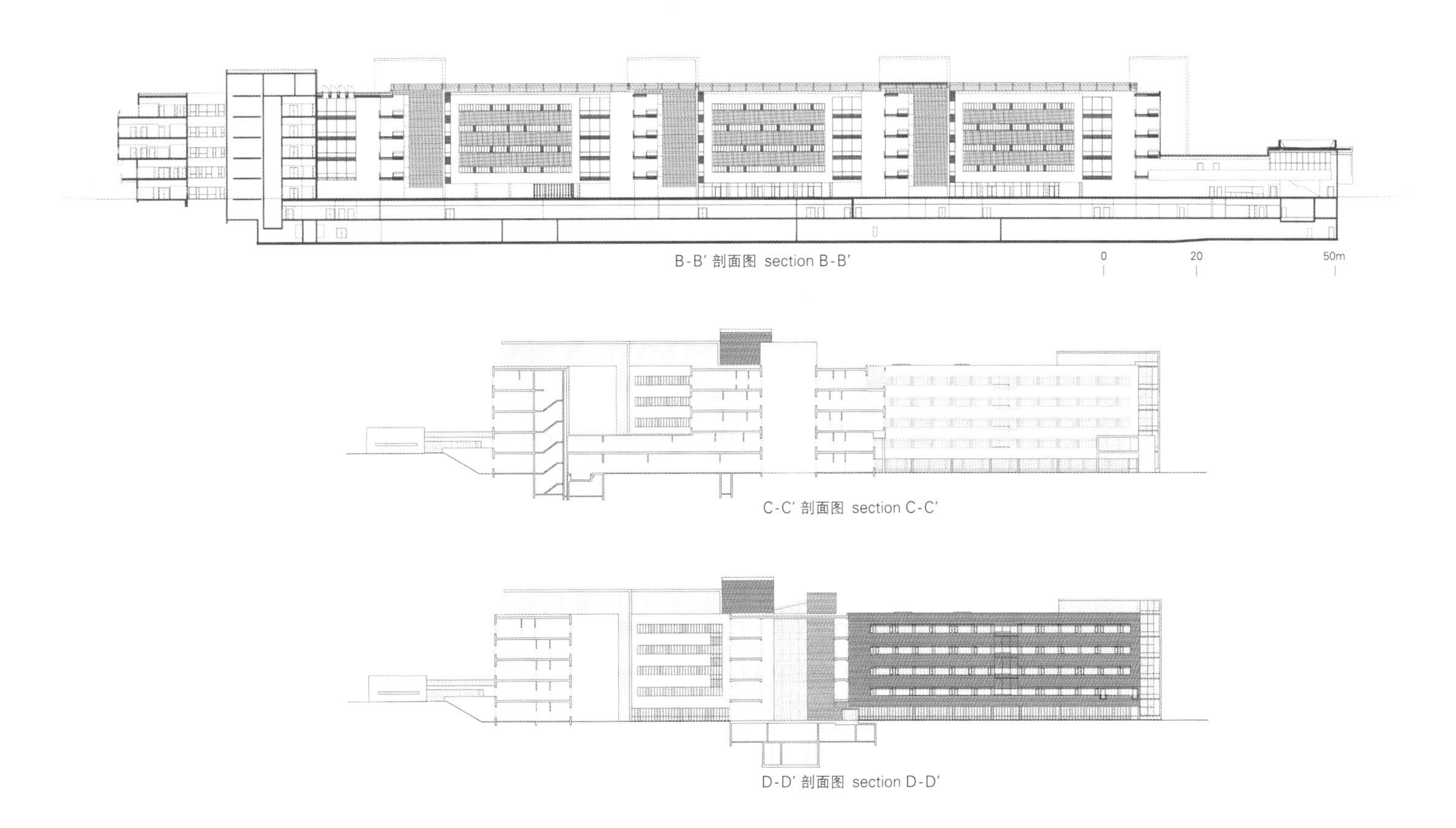

B-B′ 剖面图 section B-B′

C-C′ 剖面图 section C-C′

D-D′ 剖面图 section D-D′

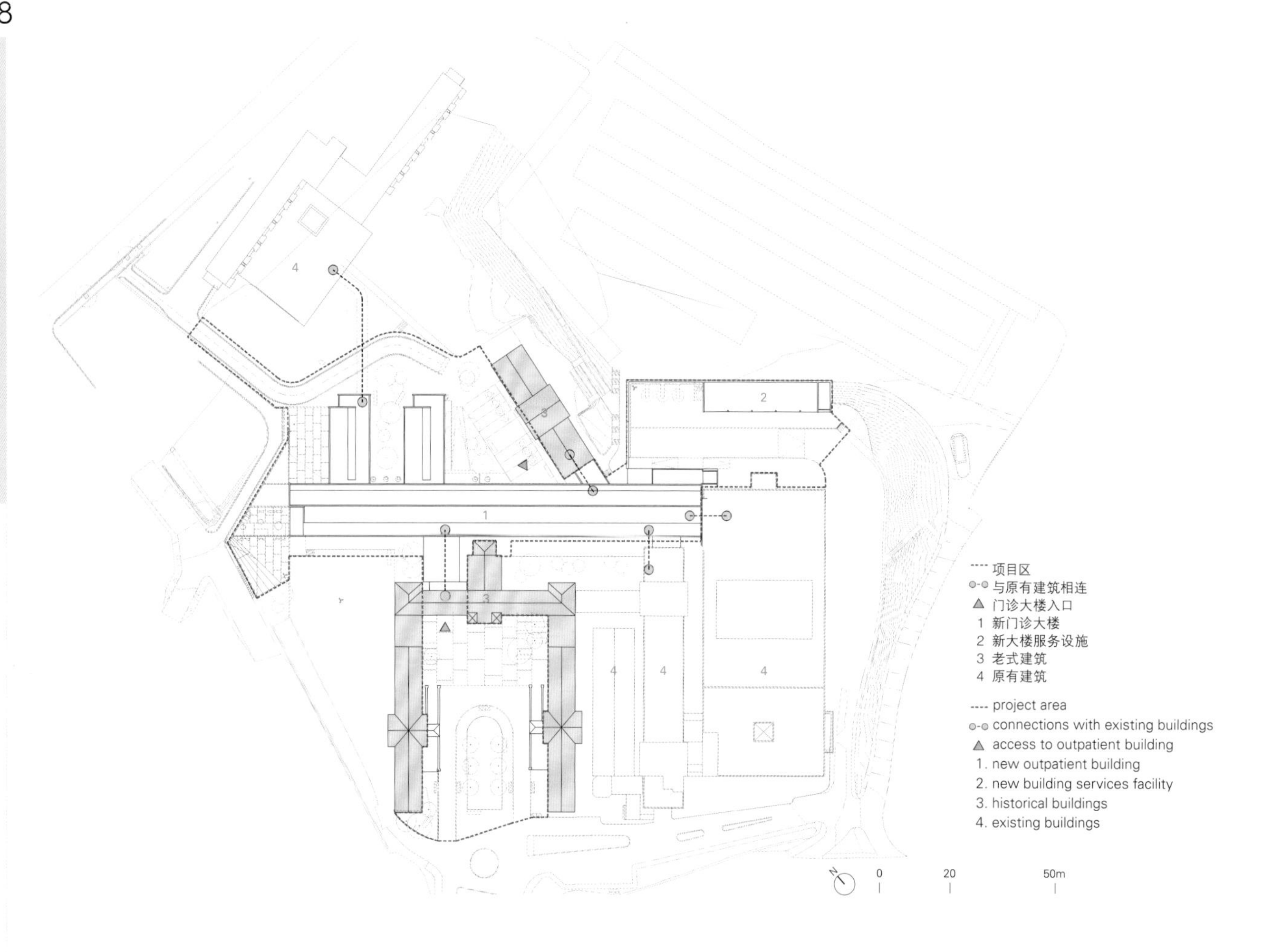

格拉诺勒斯医院扩建

Pinearq

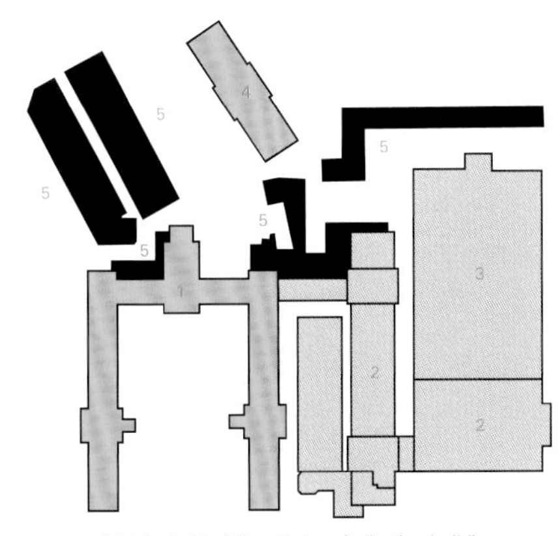

1. historical building 2. hospitalization building
3. equipment service building 4. historical home
5. buildings to be demolished
建造之前的状态
previous state before intervention

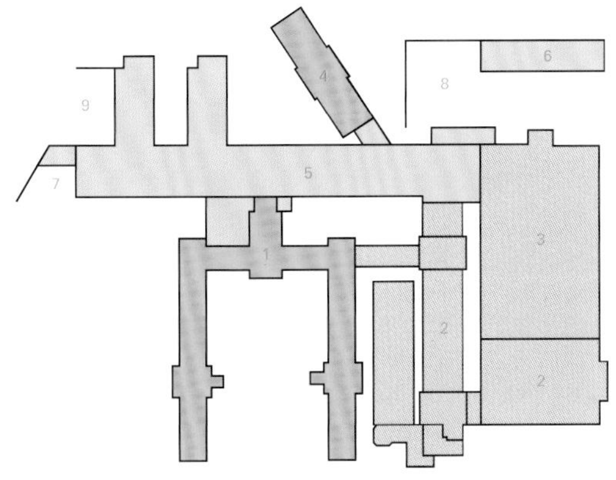

1. historical building 2. hospitalization building
3. equipment service building 4. historical home
5. new outpatient building 6. new building services facility
7. gas deposits 8. entrance of goods 9. waste exit
建造之后的状态
after the intervention

该项目包括巴塞罗那格拉诺勒斯医院新门诊大楼的扩建，以及它与原有建筑物的整合。原有医院设计于1910年，是一座历史悠久的建筑。这是现代主义建筑师Josep Maria Miró I Guibernau最杰出的作品。扩建部分立足于对原有建筑的分析，以解决与原有建筑物之间的连接问题，并增加场地面积。

新大楼是由一座作为原有建筑背景的长方形建筑（120m×15m）来确定的。在水平方向上，它依次设有急诊区、手术楼，而垂直方向上设有小教堂和老式住宅。新大楼适应倾斜的地形、地下室和一层，并允许在后部附加设置位于二层的通道。

天井区域建在新大楼和原有建筑之间，这有助于周围所有房间的通风和照明。现有的综合通道通过采用一个双层高度的大厅和一个可以俯瞰整个城市的大窗户得以扩大，这可让原有建筑物保持不变。此外，建筑师为主楼附加了两个平行的两层高的建筑，作为门诊设施的补充。干预措施还包括太平间、灭菌区、化验室、药房、辅导室、图书馆、行政文件区和综合仓库的改革。而将公共通道与技术通道区分开一直都很重要。

垂直交通设施（楼梯和电梯）经过了巧妙的设置，以加强其在原有部分中的作用。该新建筑的立面是一个无色的单一性设计，以提高原有建筑物的历史价值。

Granollers Hospital Extension

The project consists of the extension of the new outpatient buildings of Hospital of Granollers, Barcelona, and its integration with the existing building. The original existing hospital was designed in 1910 and is registered as a historic building. It was the most outstanding work of modernist architect Josep Maria Miró I Guibernau. The extension has been based on the analysis of the existing building, solving connections between the existing buildings, and also increasing areas.

The new building is defined by a rectangular building (120x15 meters) that acts as a background to the original historical building. It is defined horizontally by the emergency service, the surgical block and vertically by the chapel and the old house. The new building adapts to the sloped terrain, the basement floor and the first floor, allowing there to be an additional access through the back on the second floor.

Patio areas are created between the new building and the existing building which help ventilate and illuminate all rooms around. The present general access is enhanced with a double height hall and a large window that overlooks the city, allowing the existing building to be left unchanged. Moreover, two parallel two-story-high buildings, are added to the main building, complementing the outpatient facilities. The intervention also includes the reformation of the morgue, sterilization area, laboratory, pharmacy, coaching, library, administrative files and general warehouse. It has always been important to differentiate the public passage ways from the technical ones.

Vertical communications (stairs and lifts) are located strategically to enhance their function with existing parts. The facade of the new implementation is a neutral and homogeneous design, in order to enhance the historical value of the existing building. Pinearq

项目名称：Granollers Hospital Extension
地点：Av. Francesc Ribes S/N; Granollers, Barcelona
建筑师：Pinearq
项目建筑师：Albert de Pineda Álvarez, Javier Llambrich Bernarda
设计团队：Juan Manuel Garcia, Carles Frauca, Gerardo Solera, Pedro Pombinho
测量建筑师以及健康和安全保障人员：Enne-Gestió Activa de Proyectos S.l.p. (Imma Casado)
结构工程师：Manuel Arguijo
机械、管道和电力工程师：JG & Asociados
总承包商：Ferrovial 甲方：Fundació Francesc Ribas
建筑面积：19,500m² 施工时间：2006—2009
摄影师：©FG+SG Architectural Photography

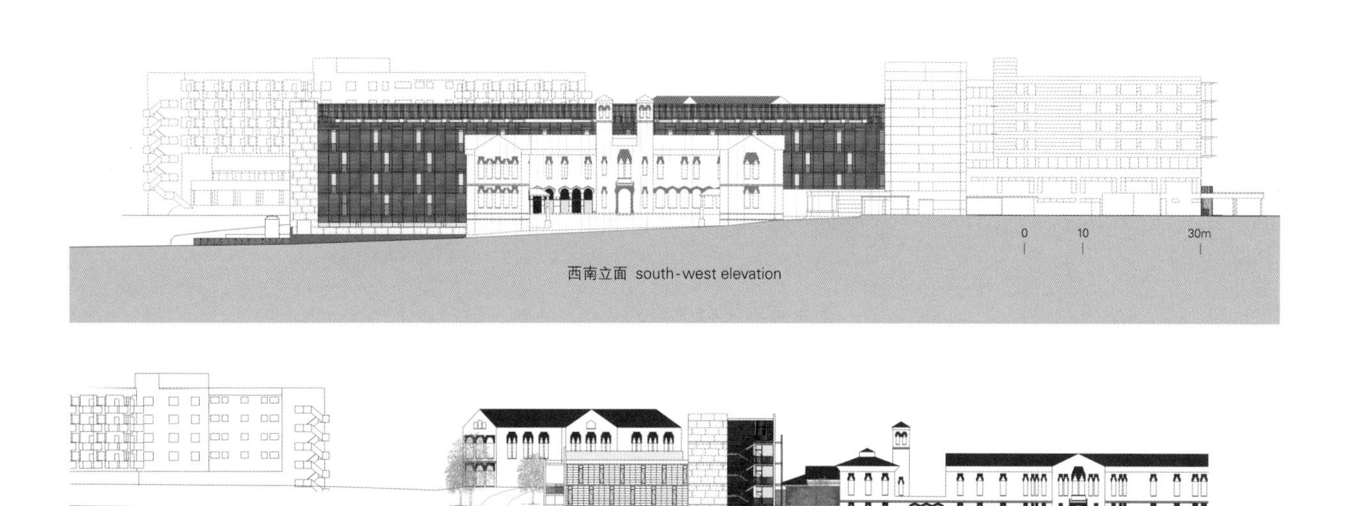

西南立面 south-west elevation

西北立面 north-west elevation

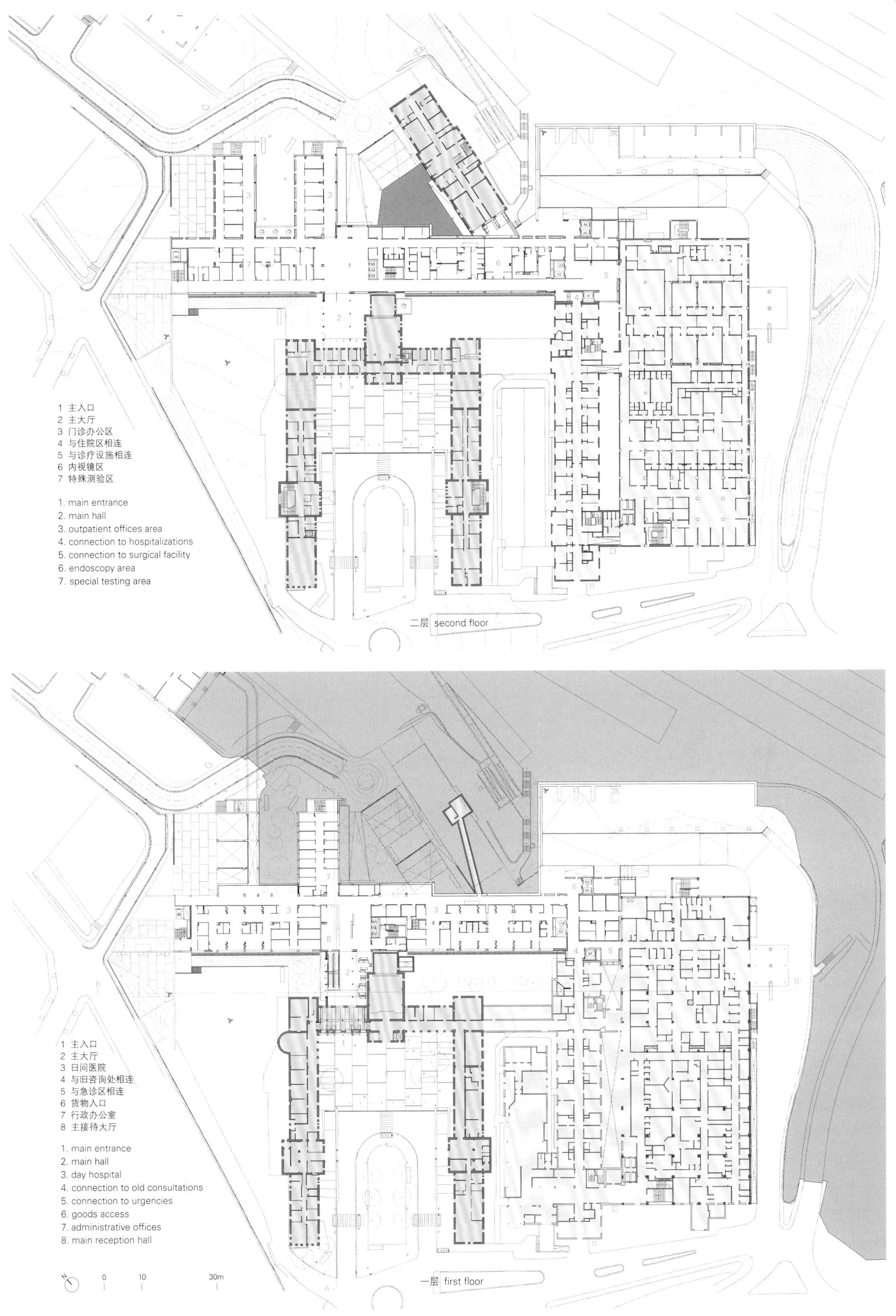
1 主入口
2 主大厅
3 门诊办公区
4 与住院区相连
5 与诊疗设施相连
6 内视镜区
7 特殊测验区
1. main entrance
2. main hall
3. outpatient offices area
4. connection to hospitalizations
5. connection to surgical facility
6. endoscopy area
7. special testing area
二层 second floor
1 主入口
2 主大厅
3 日间医院
4 与旧咨询处相连
5 与急诊区相连
6 货物入口
7 行政办公室
8 主接待大厅
1. main entrance
2. main hall
3. day hospital
4. connection to old consultations
5. connection to urgencies
6. goods access
7. administrative offices
8. main reception hall
0
10
30m
一层 first floor

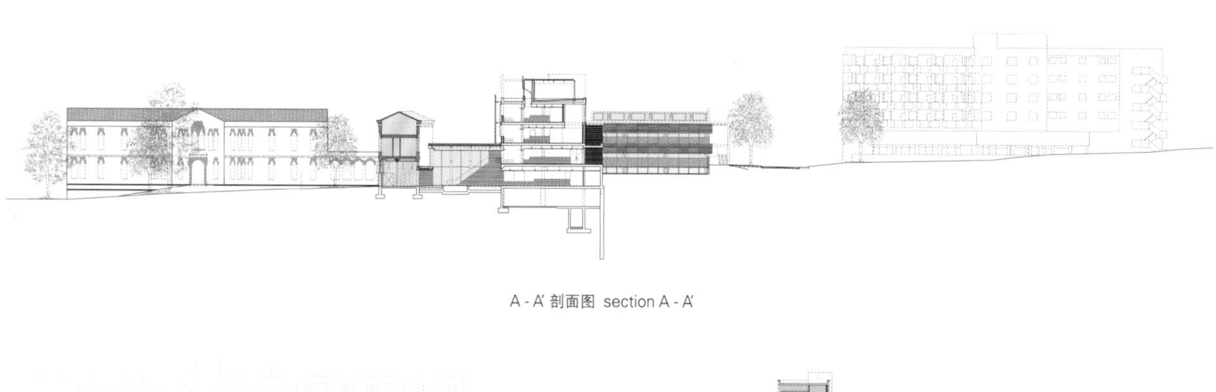

A - A' 剖面图 section A - A'

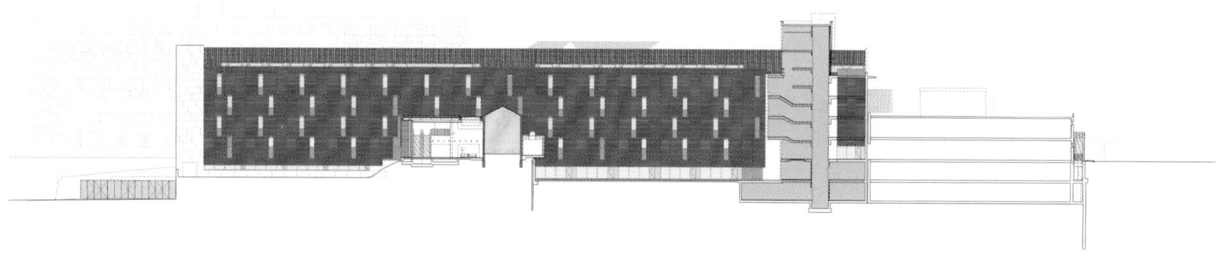

B - B' 剖面图 section B - B'

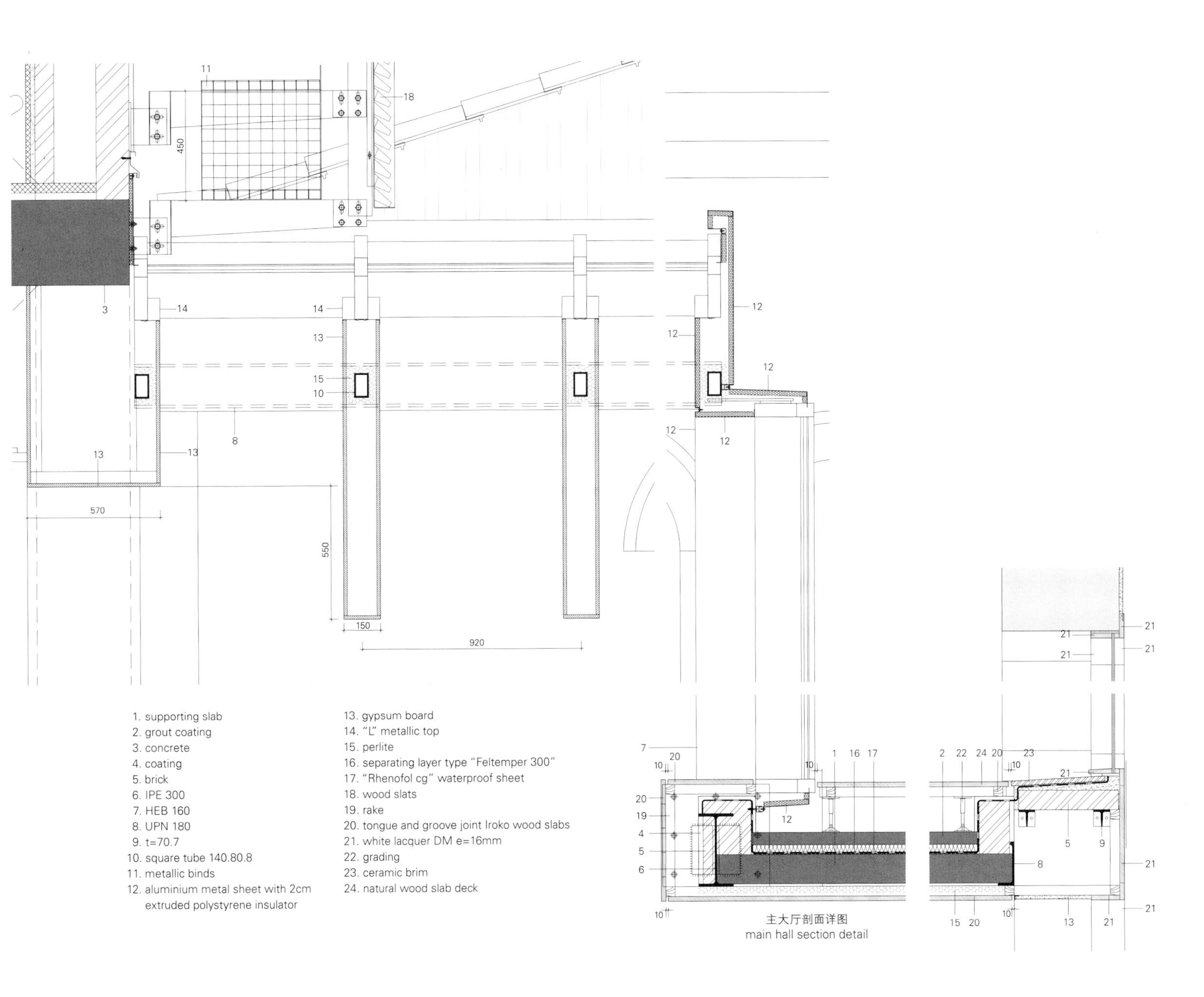

主大厅剖面详图
main hall section detail

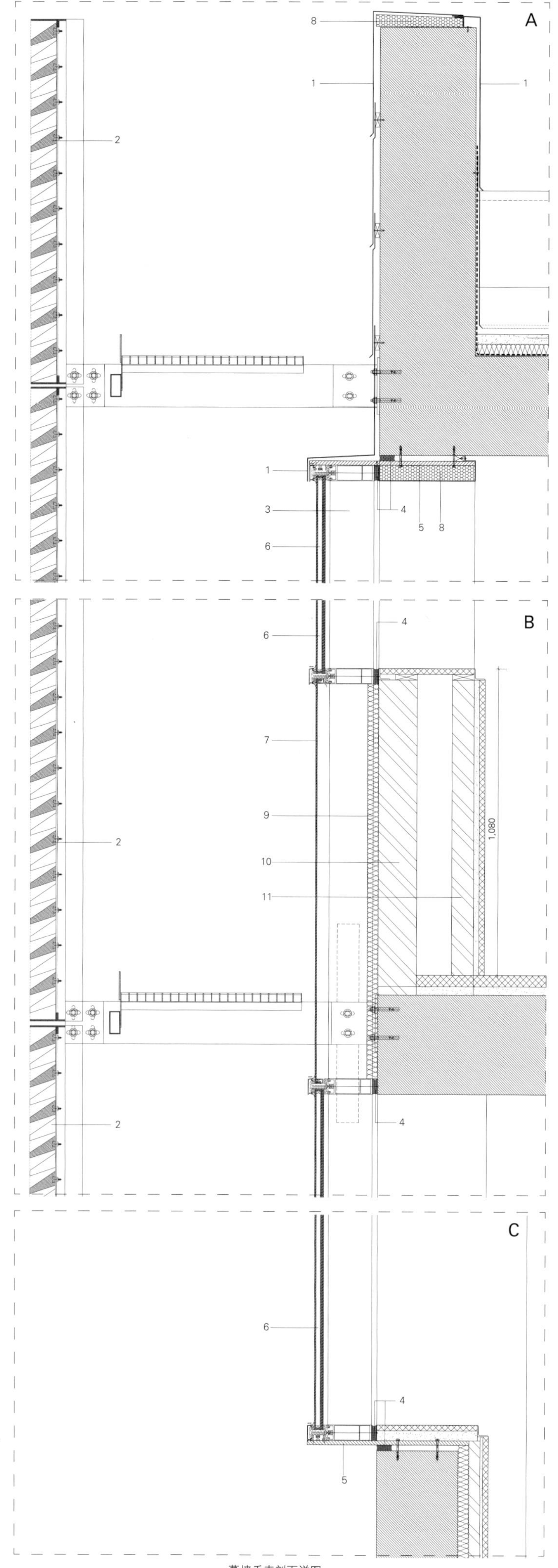

幕墙垂直剖面详图
curtain wall vertical section detail

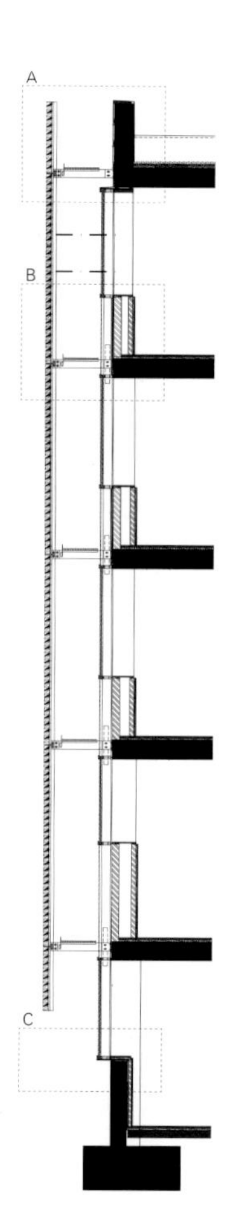

1. galvanized steel plate 2mm thickness
2. fixed panel made of treated pine slabs with frame "L" shape structure made of steel sheets 70x10mm and 60x10mm
3. curtain wall with mixed joint or equivalent with horizontal tappets
4. metallic corridor
5. galvanized steel strip e=5mm
6. safety glass with Climalit insulation 6 / 12 / 4+4
7. safety glass 4+4 with black Butiral
8. corner sheet with Alucobond or equivalent thermal insulation, with anodized aluminium e=2mm
9. extruded polyurethane e=4cm
10. ceramic brick
11. large format ceramic tongue and groove joint with plaster coating

先进医疗技术研究所——塞维利亚的医院

MGM Arquitectos

人们习惯了城市的形式，城市的每个组成部分都是单独设计的，也就是说，有人追溯城市的形式在先，然后建筑物才得以设置到位。这至少意味着一个问题：建筑与城市的融合非常艰难。

该项目在如下信念指导下进行：由于未来的大规模建筑，它必须作为城市本身的一部分来安置。因此，这意味着该项目还包括建造一条横穿建筑的宽阔走廊；建立一座大厅，以连接该医院的各个复杂部分；同时这条走廊必须在不同楼层上配置不同的门廊，如面向城市和乡村风景的巧妙的凸窗。这些门廊将医院病房部分和病人家属与医生的休息区组织到一起。该建筑还包含有调和建筑主体和空间的治疗凸窗，使人们看到历史悠久的城市和当代城市各个复杂部分的融合。

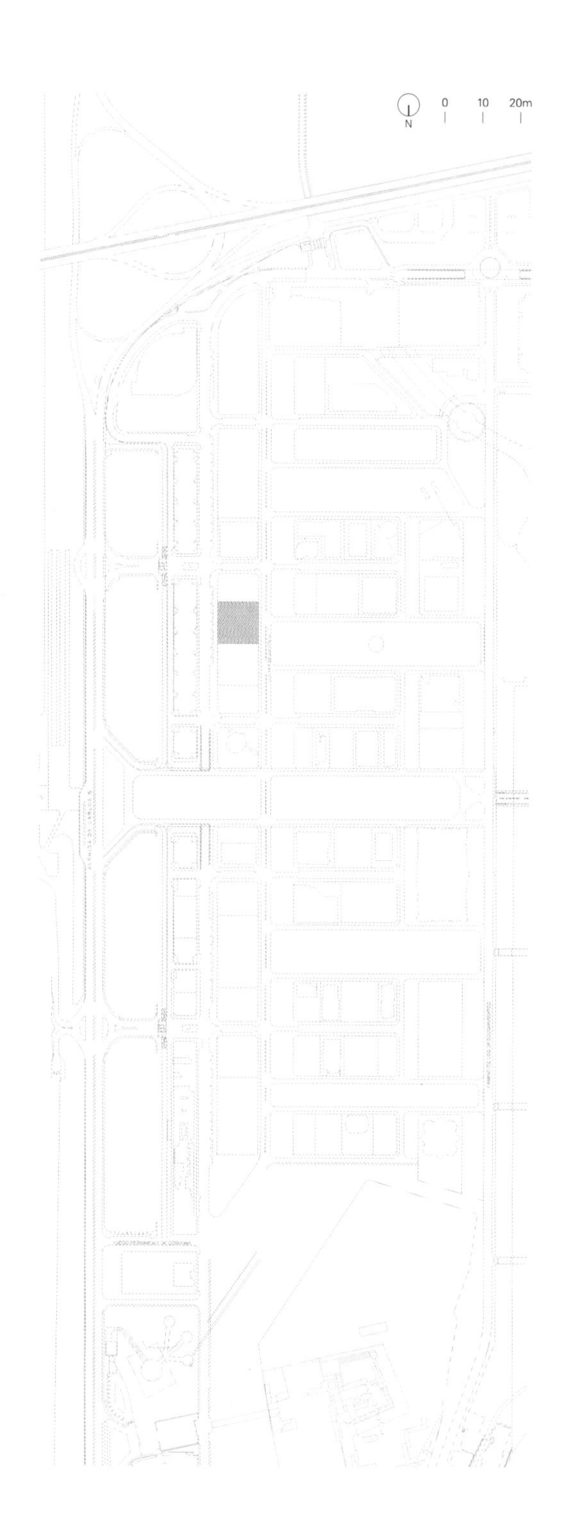

至于它的运作方式：功能程序的合理性和内部空间的有益性与友善性在大楼内统一起来。为此，建筑师提出了一个非常合理的平面图，它位于包括四个垂直交通设施核心的朝向南面的功能组合带的结构中。该模式的主要优点是缩短了交通流线，使得公共与私人空间分明、各房间相互独立（总是朝南方向），并且使建筑的交通流线大部分处于水平层而非垂直方向。

楼层平面是围绕三个纵向天井组织的，其中一个将构成建筑物的主入口，而其他两个构成设有患者房间的楼层布局。主入口区与建于其中的新走廊连接起来，就像一个巨大的露天门廊，充满了喧嚣和噪音：该空间的入口受到控制，在该空间，健康中心外的人都还没有进入医院封闭的空调区域。可以说，这个"过滤街"同时也是构成医院两个大型入口的外部和内部空间。

建筑师意识到，时间的推移、技术的进步和服务的需求迫使这些建筑物被改建。尽管这些情况不是完全如人所愿的，但新医院的提案不可避免地考虑了这些情况，所以他们认为，所提出的楼层规划是非常合理有效的。除了交通流线的合理性和患者的亲切感，所有房间都朝向南面也是主要特征之一。建筑师认为，借助大楼的南北朝向采用自然治疗（在这种情况下是指太阳的方位）伴随药物治疗的方法是可取的。

Institute of Advanced Medical Techniques – Hospital in Seville

We are accustomed to the form of the city and what constitutes it is negotiated separately; that is to say, someone traces the form of the urban and then the buildings are set in place. This presupposes at least one problem: only with difficulty are buildings imbricated with the city.

The project proceeds from the conviction that due to the large size of the future building, it must be posited as a part of the city itself. As a result, this means proposing a work that includes a wide street which goes through the building, creating a city hall, from which to connect the whole complex program of this hospital; at the same time this street would have to collocate different vestibules at different levels, like strategic miradors facing the urban and rural landscape. These vestibules organize the program of hospital rooms and the relaxation of family members and doctors. Therapeutic miradors are involved that reconcile bodies and spaces and propose a complicity of the gaze intersecting with the historic city and the complexity of programs of the contemporary city.

As for its way of working: united in the building are the rationality of the functional program and the helpfulness and amiability of the interior spaces. To that end we have proposed a highly rational

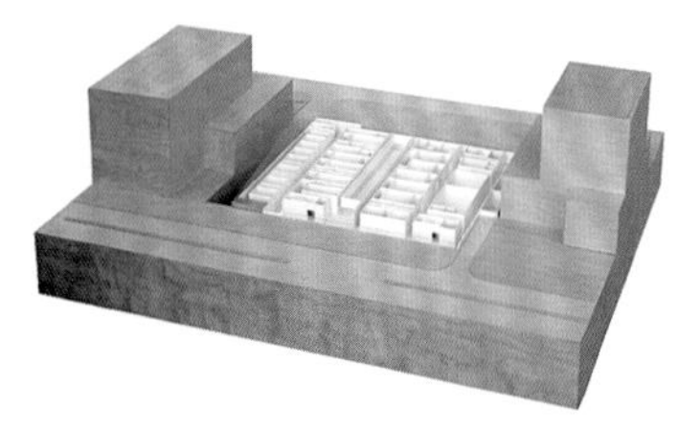

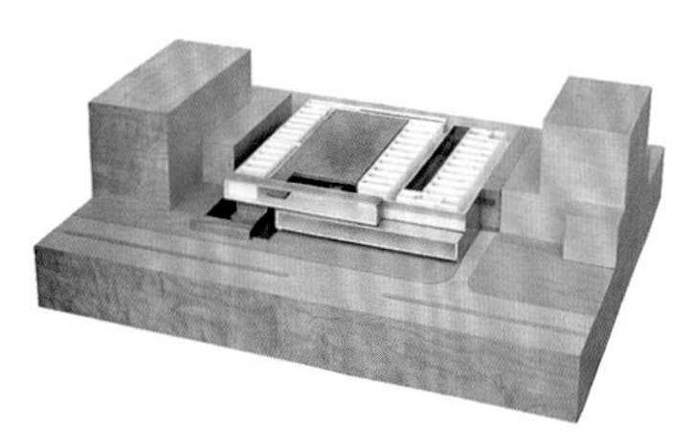
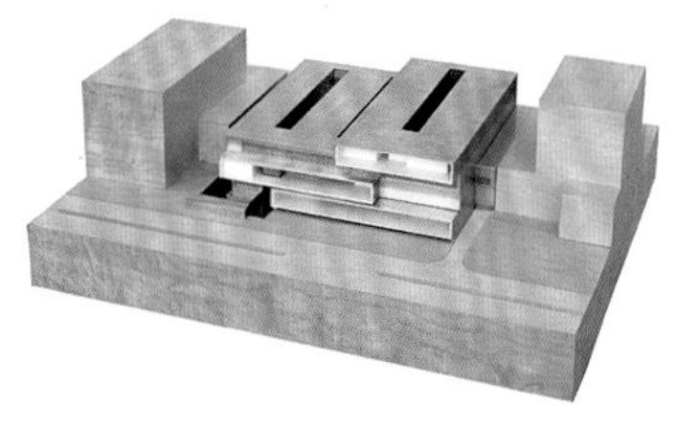

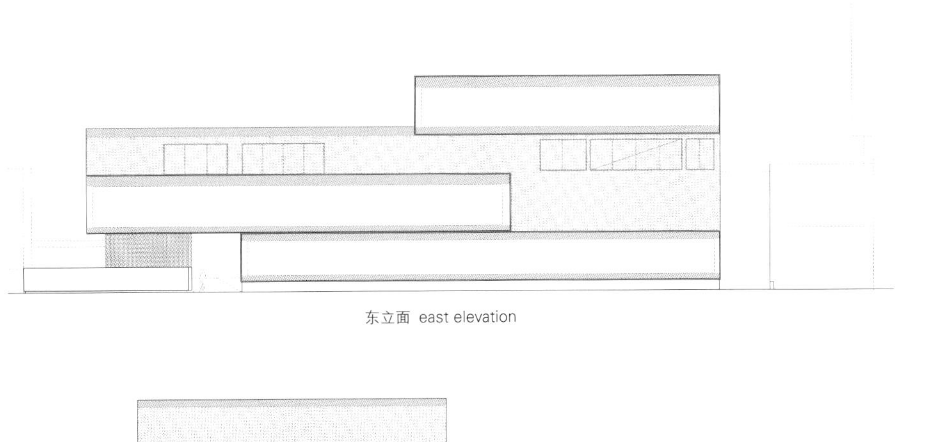

东立面 east elevation

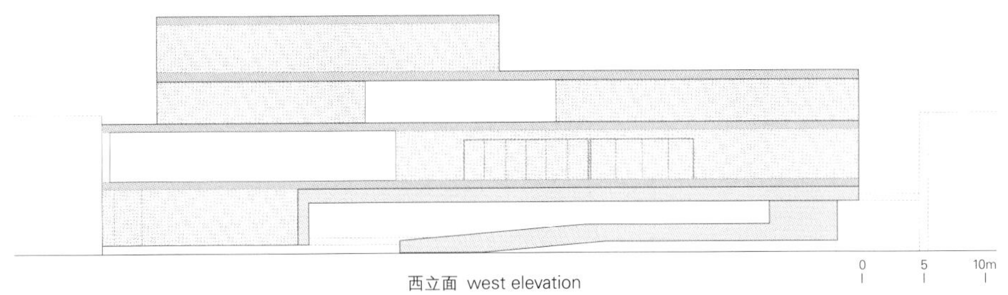

西立面 west elevation

项目名称：Institute of Advanced Medical Techniques – Hospital in Seville
地点：Science and technology park Cartuja 93 Seville, Spain
建筑师：José Morales, Sara de Giles
合作者：Ángel Barreno, Carlos Bauzá, Jordi Bolaños, Jairo Fernández, Lola Hermosilla, Isabel Jiménez
服务方：Alberto Germá 模型：Pablo Olias, Jaime Hernando
建筑模型：Giulia Barra, Elena Jiménez, Juan J. Olmo, Rubén Olivares
建筑推广：A.G.E.S.A. Empresa Pública de Gestión de Activos S.A.
项目和场地管理：A.Y.N.O.V.A. SA
建筑面积：14,123m² 造价：EUR 14,663.349
施工时间：2009—2012
摄影师：©Jésus Granada

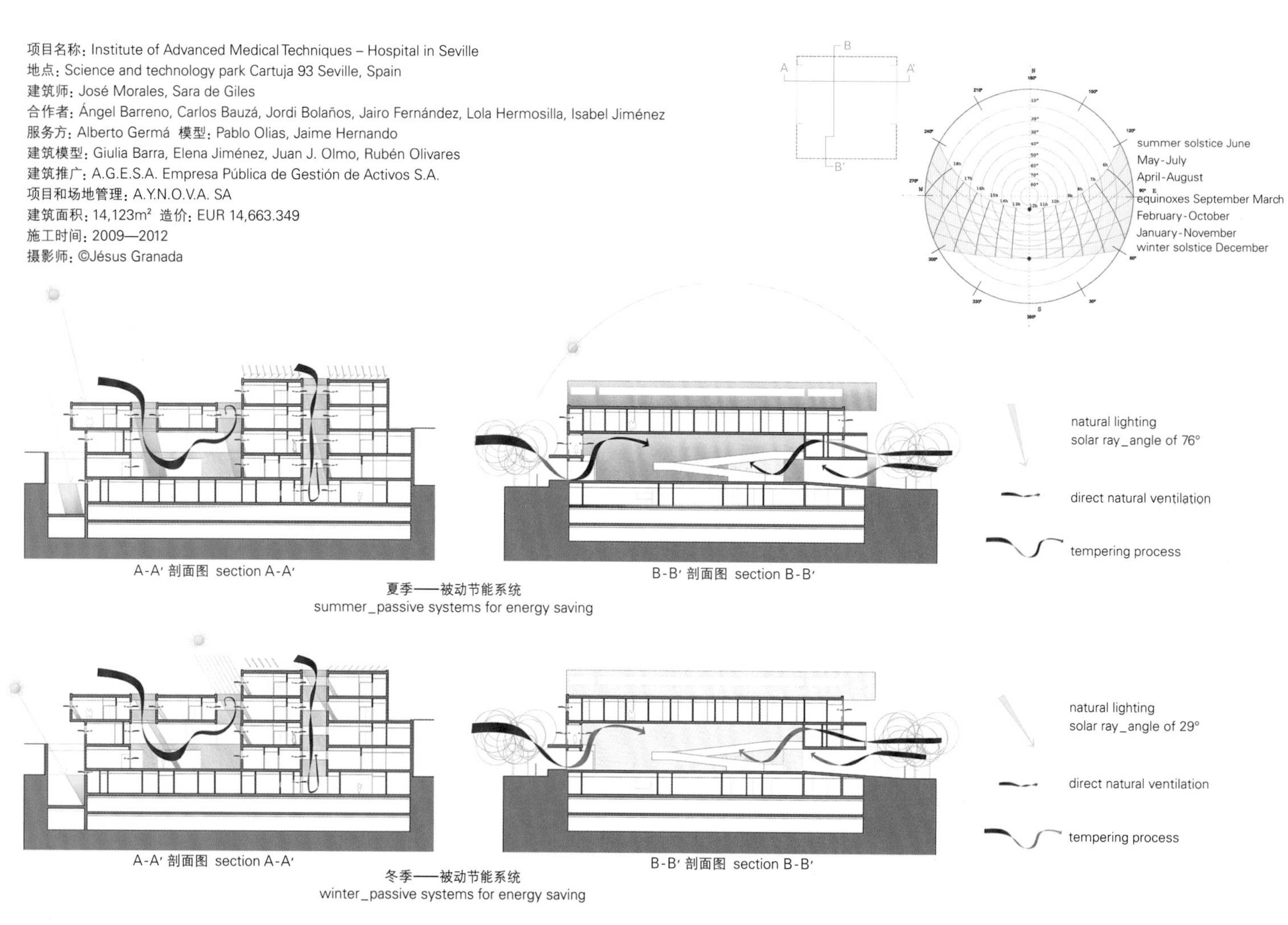

A-A' 剖面图 section A-A'
B-B' 剖面图 section B-B'
夏季——被动节能系统
summer_passive systems for energy saving

A-A' 剖面图 section A-A'
B-B' 剖面图 section B-B'
冬季——被动节能系统
winter_passive systems for energy saving

1 行政管理区
2 急诊咨询处
3 观察室
4 员工起居室1
5 大厅
6 行政管理处
7 清洁室
8 厨房
9 投影室和会议室
10 主入口
11 等候区
12 接待处
13 咖啡厅
14 远程放射室
15 X线断层摄影室1
16 伽马成像
17 员工起居室2
18 手术室
19 候诊室
20 服务主管办公室
21 超声波检查处
22 乳房X线照相术
23 磁共振室
24 康复室
25 放射科1
26 全影X线诊疗室
27 X线断层摄影室2
28 放射科2
29 实验室
30 药品储藏室
31 多功能咨询处
32 房间类型
33 天井
34 等候室咨询处
35 家庭聚集区和休息区

1. administrative area
2. emergency consultation
3. observation room
4. living room for staff 1
5. hall
6. administration
7. clean office
8. kitchen
9. projection room and meetings
10. main entrance
11. waiting area
12. reception
13. cafe
14. teleradiology
15. tomography 1
16. gamma camera
17. living room for staff 2
18. operating room
19. waiting room
20. office of the manager of sevice
21. ultrasonography
22. mammography
23. megnetic reconance
24. recovery room
25. radiology 1
26. orthopantomography
27. tomography 2
28. radiology 2
29. laboratory
30. medicines stock
31. multipurpose consultation
32. room type
33. patio
34. waiting room consultation
35. family meeting area and rest

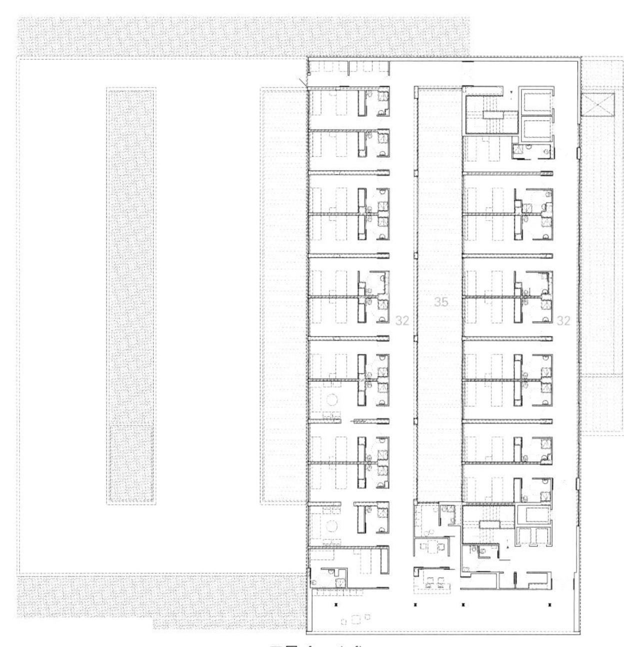
四层 fourth floor

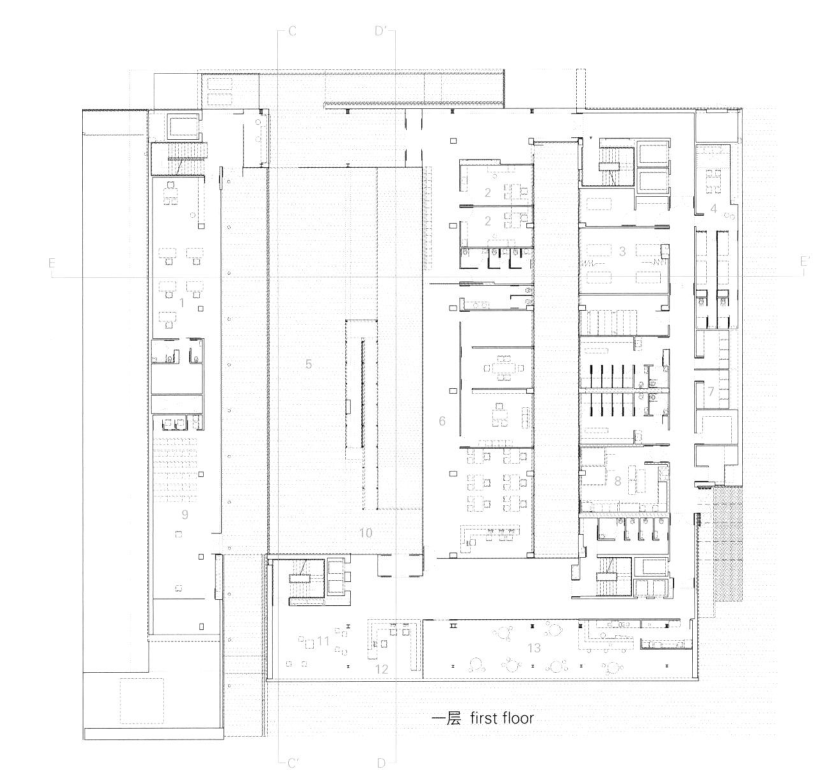
一层 first floor

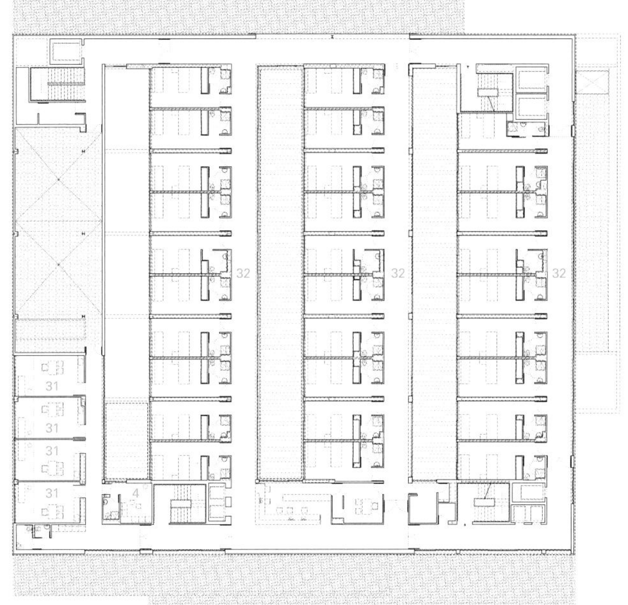
三层 third floor

地下一层 first floor below ground

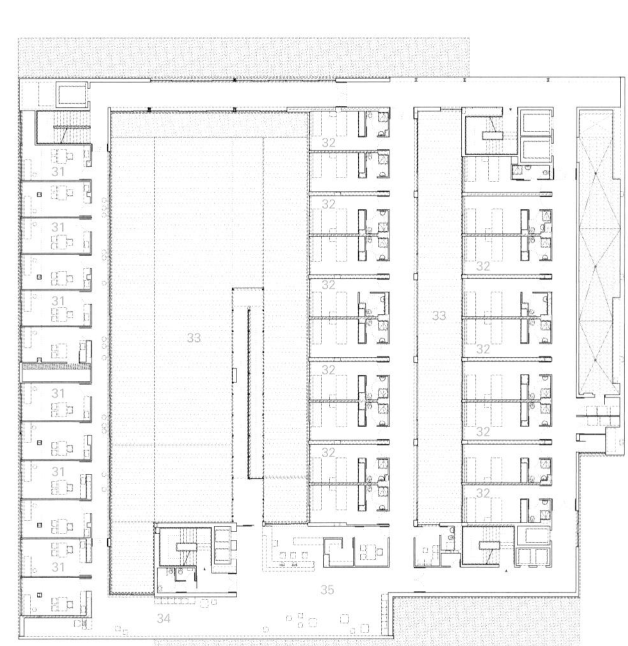
二层 second floor

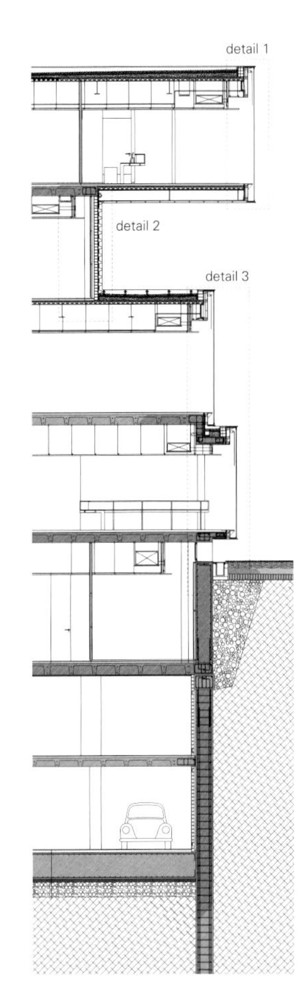

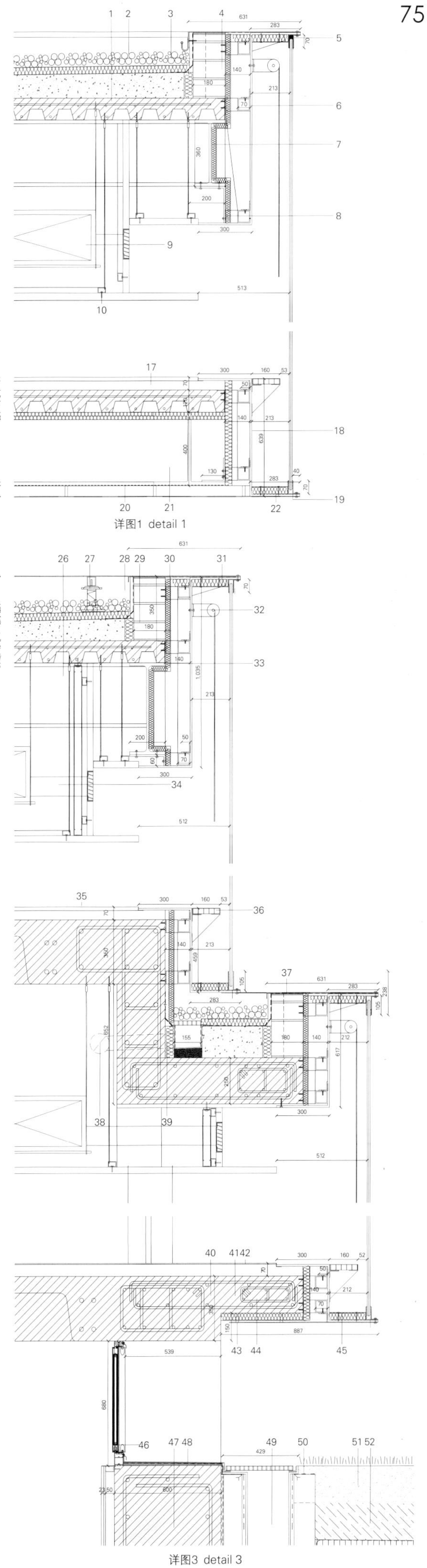

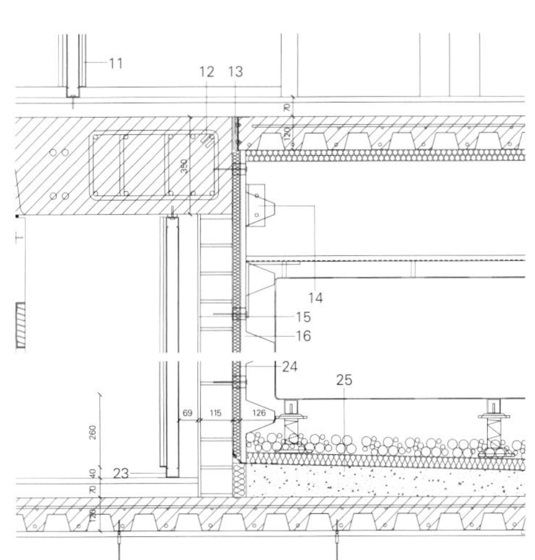

1. slab metal consisting of: steel deck and compression coat. thick=12cm **2.** protective layer of gravel. pebble, maximum diameter 25 mm, with a minimum thickness of 50 mm **3.** plastered type mortar monolayer, color concrete, thick=2cm **4.** perforated brick parapet **5.** I profile glass clamp, 70.40.5 mm **6.** tubular substructure 70.70.5mm, steel, galvanized, screwed sheets to forged or structural element attachment **7.** laminated steel structure. standard secondary profile IPE/HEB **8.** I profile galvanized steel screwed into secondary profile wings **9.** grid impulse **10.** guides ceiling mounting. galvanized steel anchored by threaded rods forged M4 **11.** gypsum board 1 side treated with melamine, set on dry wall, total thickness:12.5mm **12.** reticular reinforced concrete slab, thick=30cm(20+10) **13.** union between different types of floor slabs by elastic material thick=2~3 cm **14.** armed beam stiffeners **15.** perforated brick wall **16.** air chamber **17.** levelling mortar, e=5cm, type HM/20/B/40/I **18.** structural sheets of galvanized steel, thick=6mm **19.** filled with elastic material type neoprene or similar **20.** mesh base against birds attached to the substructure, thick=15 mm **21.** reinforced concrete beam 40cm **22.** thermal insulation, fiberglass thick=40mm **23.** aluminum L-profile galvanized, 40.58.3 mm **24.** thermal insulation: projected polyurethane. thick=30 mm **25.** slope formation of M-2 mortar(1:8). slope of 1% to 5%, minimum thickness 5cm **26.** reinforced concrete beam 80cm. with reduced end for installations **27.** adjustable plots. located in parallel lines every 75cm **28.** lifeline: stainless steel AISI 316, 8mm diameter cable **29.** junta rellena de material elastico para permitir dilatacion horizontal. thick=5cm **30.** profiled sheet "L" shaped galvanized steel 120.50.3mm **31.** galvanized steel to high temperature 200.80mm each 150cm **32.** laminated motorized roller shade **33.** projected thermal insulation, thick=30mm **34.** air conditioning ducts **35.** terrazzo micrograin white, 40x40 cm tiles placed on a bed of mortar, e =2cm **36.** tramex 15.15. 3mm hot dip galvanized **37.** hot-dip galvanized steel sheet e=3mm, top cover, folded **38.** hanging rainwater collector **39.** embossed aluminum sheet in color, dim:2,000, 1,000mm. thick=4mm, screwed to battens **40.** reticular reinforced concrete slab, thick=30cm (20+10) **41.** reinforced concrete slab floor, thick 25cm **42.** continuous pavement on base epoxy mortar regularized with antibacterial treatment and impact resistant **43.** pins of galvanized aluminum **44.** reinforced concrete slab, thick=25 cm **45.** self-protected aluminum foil attached to sheet, thick=1.5mm **46.** fixed anodized aluminum carpentry in color, with 1 sliding placing on the end **47.** reinforced concrete wall **48.** aluminum sheet asphalt protection **49.** precast concrete gutter **50.** vent galvanized aluminum, put every 4m **51.** mantle of topsoil. minimum thickness: 25cm **52.** compacted soil layer, 20cm minimum thickness

floor plan structured in bands of south-facing functional packages including four vertical communication cores. The main advantages of this scheme are the shortening of trajectories, allowing public-private distinction, the isolation of the rooms (always south orientation) and the predominance of the building's horizontality as against vertical communication.

The floor plan is organized around three longitudinal patios, one of which will structure the building's general entrances, and the other two the organization of the floors with patients' rooms on them. The main entrance area connects up with the new street created in its interior, which, like a great open-air vestibule, is reserved for bustle and noise: a space with controlled access in which people from outside the health center have still not entered the closed, air-conditioned area of the hospital. It could be said that this "filter street" is at once an external and internal space that structures the two great access points to the hospital.

We are all aware that the passing of time, the advance of technology and the demand for services oblige these kinds of buildings to be remodelled. Even though these circumstances are not wholly desirable, it is inevitable that the proposal for a newly built hospital considers such circumstances, and so we consider that the rationality of the floor plan proposed is highly efficient. That all the rooms are south-facing has been one of the chief wagers, in addition to the rationality of the trajectories and the intimacy of the patients. We think it advisable to accompany medical therapy with that of nature (in this case the position of the sun) thanks to the north-south orientation of the building. MGM Arquitectos

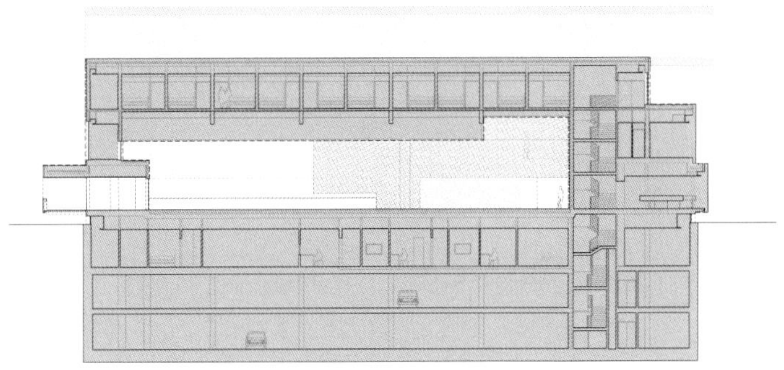

C-C' 剖面图 section C-C'

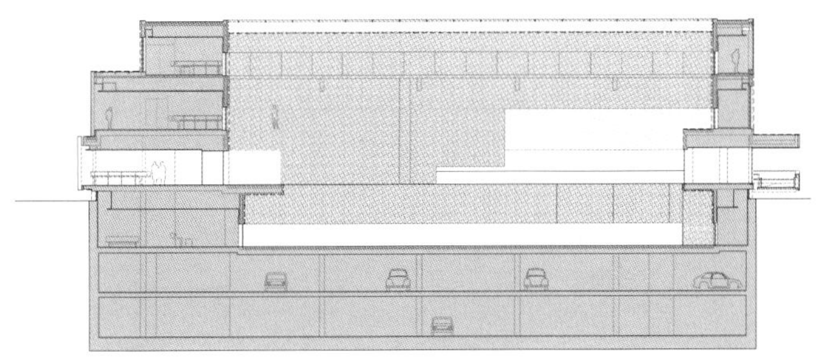

D-D' 剖面图 section D-D'

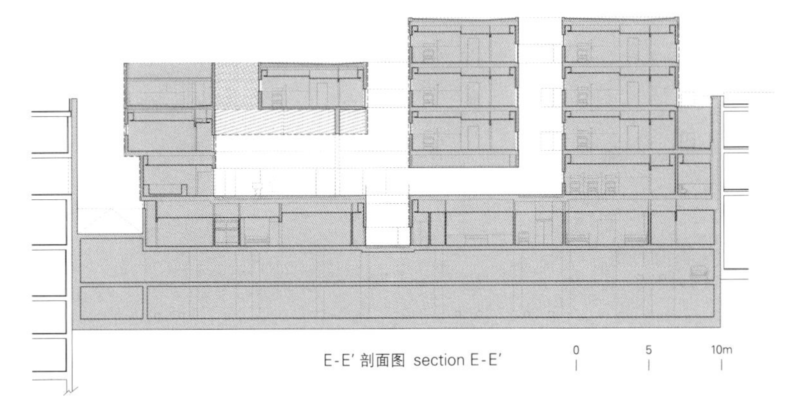

E-E' 剖面图 section E-E'

胡安卡洛斯国王医院

Rafael de La-Hoz Arquitectos

Hospital Rey Juan Carlos

Hospital Rey Juan Carlos

建筑师提议将市民转变为这座新型医院的消费者，这样做除了辅助我们公认有效的的医保系统，还可以使人们感觉一直处于关怀的中心，受到全面的关注。

这座新医院模型的配置主要考虑三个基本因素：效率、采光和宁静。它们是医疗建筑最重要的因素，也是住宅建筑最重要的因素。从概念上来说，这座新型医院是在基座上进行布局的，而该基座将医保单元、门诊和诊疗部门组织了起来。医院结构分为三个模块，或说几座平行的建筑，反映了医院结构所能具有的最佳特征：灵活、开阔、功能清晰且具有水平的交通流线。这一结构上面布局了两个住院治疗构件，它们是两个椭圆形顶冠，带有和缓的曲线，让人们看到与给人沉闷感的理性主义"条状住宅建筑"不同的建筑形式，并且吸收了近期住宅建筑的最优点：拆除走廊从而消除令人烦扰的噪音、集中式交通，以及一个普通中庭周围具有的采光条件和静谧。

整体设计理念是基于建造一座具有如下特征的医院建筑方案：它应该满足项目和预期资金的需求，也应该尝试以一种现代且具有吸引力的建筑形式来呼应项目复杂的功能性。此外它还特殊考虑了建筑规模、遮阳设备，以及将病房在整个医院中凸显出来这一最重要的方面。建筑设计在一座大型基座上布局，该基座上有医院的多种医务区，还有两座玻璃体量，内设病房。这种建筑手法在组成基座的三个棱柱内产生了一种复杂而集中的空间系统。三个棱柱相互调整适应，像一架"治疗机器"。

基座除了进行功能性处理，还进行了象征性处理，塔楼便是明证。病人活动空间按照需求进行规划，以找到使他们更好地感受协调性和阳光的最佳方式。该建筑在各个区域之间建立起了完美的功能性联系，并为内外部空间的设计提供了最大灵活性。将内外交通流线区分开，并区分垂直交通流线核心筒的入口和特征，使建筑方案得以阐明。

两座塔楼的位置与功能性需求相关，它们有一条通道可以直达手术室、产房、急诊室和诊疗室。建筑的功能性与病人和其家属在医院期间的生活方式相呼应。建筑策略是将项目中包含的所有元素进行布局。基座中包含的三个棱柱将交通流线以一种典型方式进行排布。边缘的两个棱柱排布的方式使一个适于内部使用，而另一个适于外部使用。当内外部交通流线繁忙的时候，中间的体量就可供共同使用。

建筑师建造这座建筑的重点之一是可持续性，他们考虑了太阳方位的状况、地形、建筑环境、周围植被，也没有忘记考虑城市适用条件。环保材料和可再生能源科技也融入到建筑中来，以节约资源并优化运行成本，同时通过绿化屋顶来为建筑内部提供自然采光和通风。

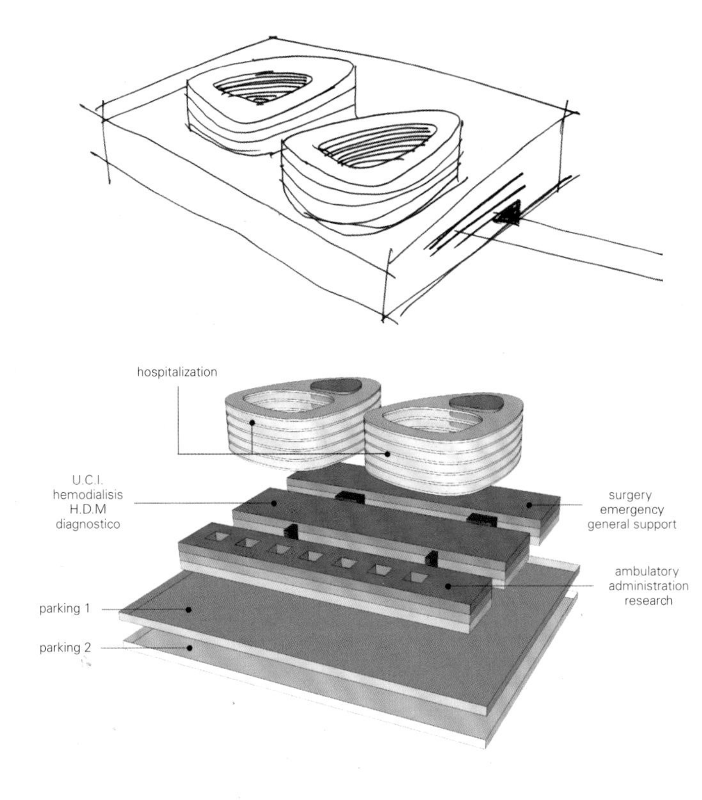

Rey Juan Carlos Hospital

We propose to transform the citizen into customer, for a new type of hospital, which in addition assists the proven effectiveness of our healthcare system, so that they can feel at all times the center of all care, giving them all attention.

This new hospital model is configured in three basic elements: efficiency, light and silence, which are the best about hospital architecture and the best in residential architecture.

Conceptually, the new hospital is arranged on base that gives structure to the healthcare units, outpatient diagnosis and treatment. Structured in three modules or parallel buildings that reflect the best hospital's main structures: flexibility, expansion, functional clarity and horizontal circulations.

On this structure are arranged two units of hospitalization, so two oval crowns with gentle curves drawn give a different view from the depressive sensory residential forms of the rationalist "block bar", and draw on the best of recent residential architecture: the elimination of corridors and in consequence the elimination of annoying noise, concentric circulation, light and silence around a common atrium.

The overall concept is based on the architectural proposal for a hospital of this nature; it should allow adaptation to the requirements of the program needs, and expected financial requirements. It has also sought to respond to the complex functional program with contemporary and attractive architecture. It has had particular regard to the human scale, solar protection and above all to distinct the patients' spaces through the hospital.

The design is structured on a large base, which encompasses various medical areas of the hospital, holding two volumes of glass where the ward is developed. This approach develops to a complex system of articulated spaces within three prisms that make up the base, geared to each other as if it were a machine, a "healing machine".

In addition to the functional treatment of the base, a symbolic treatment was essential to remark the towers. The space of the patients is planned only thinking in the needs and in the best way for them to feel better with harmony and light.

项目名称：Rey Juan Carlos Hospital
地点：Gladiolo st. s/n 28933 Móstoles, Madrid, Spain
建筑师：Rafael de La-Hoz Castanys
建筑设计师：Hugo Berenguer, Francisco Arévalo, Miguel Maíza, Jacobo Ordás, Carolina Fernández, Encarna Sánchez, Gonzalo Robles, Javier Gómez, Ignacio Jaso
项目基础：Miguel Maíza, Jacobo Ordás, Gonzalo Robles, Javier Gómez
项目执行：Miguel Maíza, Sigfried Burguer, Hugo Berenguer, Jacobo Ordás, Gonzalo Robles, Peter Germann, Laura Díaz, Fernando de la Fuente, Saúl Castellanos, Carmen Salinas, Ignacio Jaso
项目协作：Miguel Maíza, Jacobo Ordás
项目协作合作者：Fernando de la Fuente
电脑绘图设计：Luis Muñoz, Daniel Roris
立面：Permasteelisa, Cricursa, FERGA, Doval Building
用地面积：90,000m²
地上建筑面积：69,782m²
地下建筑面积：24,923.20m²
总建筑面积：94,705.49m²
竣工时间：2012
摄影师：©Duccio Malagamba(courtesy of the architect) - p.80~81[bottom], p.84~85, p.86, p.87, p.89[bottom], p.91, p.92~93
©TAFYR, S.L.(courtesy of the Madrilenian Public Health Service) - p.78~79, p.99(except as noted)
©Alfonso Quiroga - p.80~81[top], p.89[top]

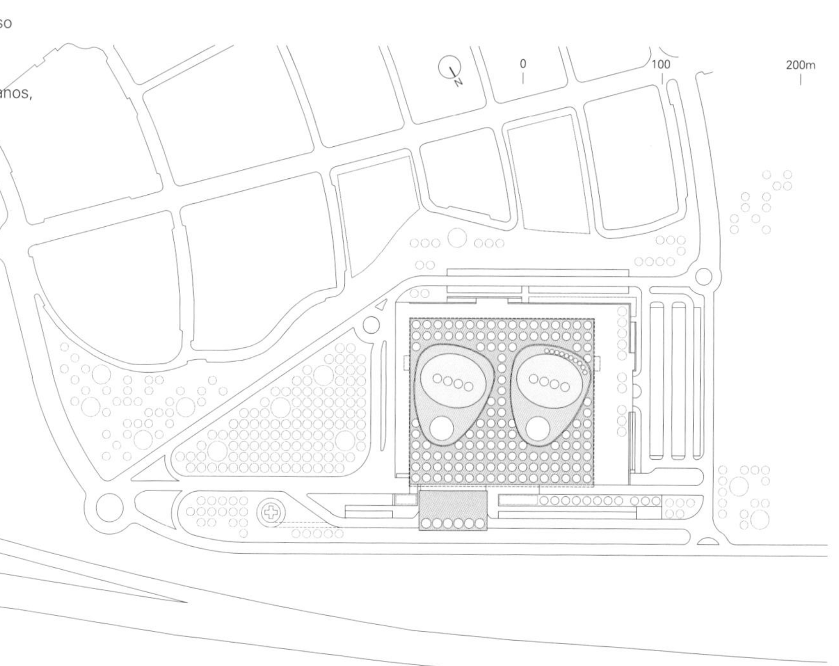

©Obrascon-Huarte-Lain(courtesy of the architect)

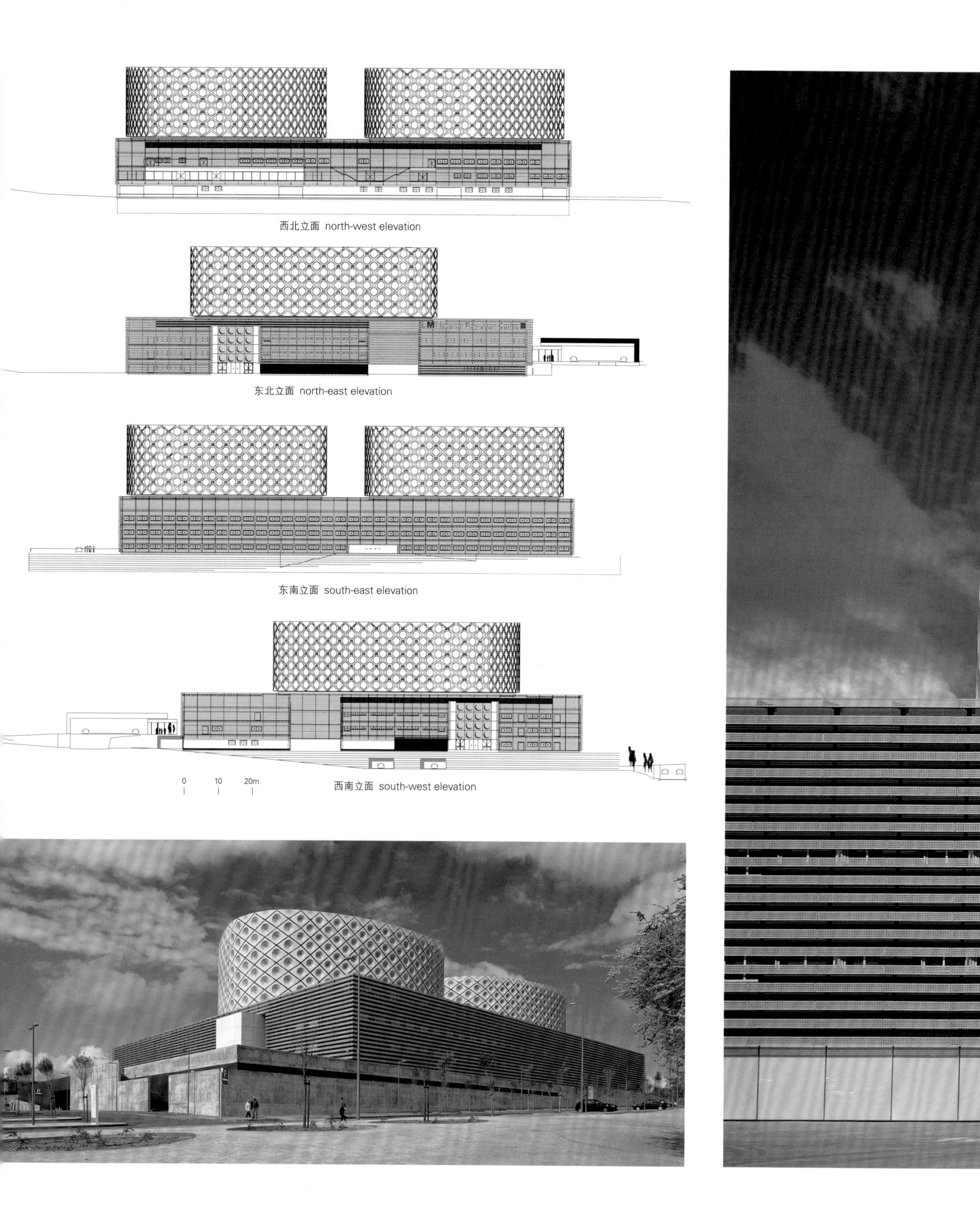

西北立面 north-west elevation

东北立面 north-east elevation

东南立面 south-east elevation

0 10 20m

西南立面 south-west elevation

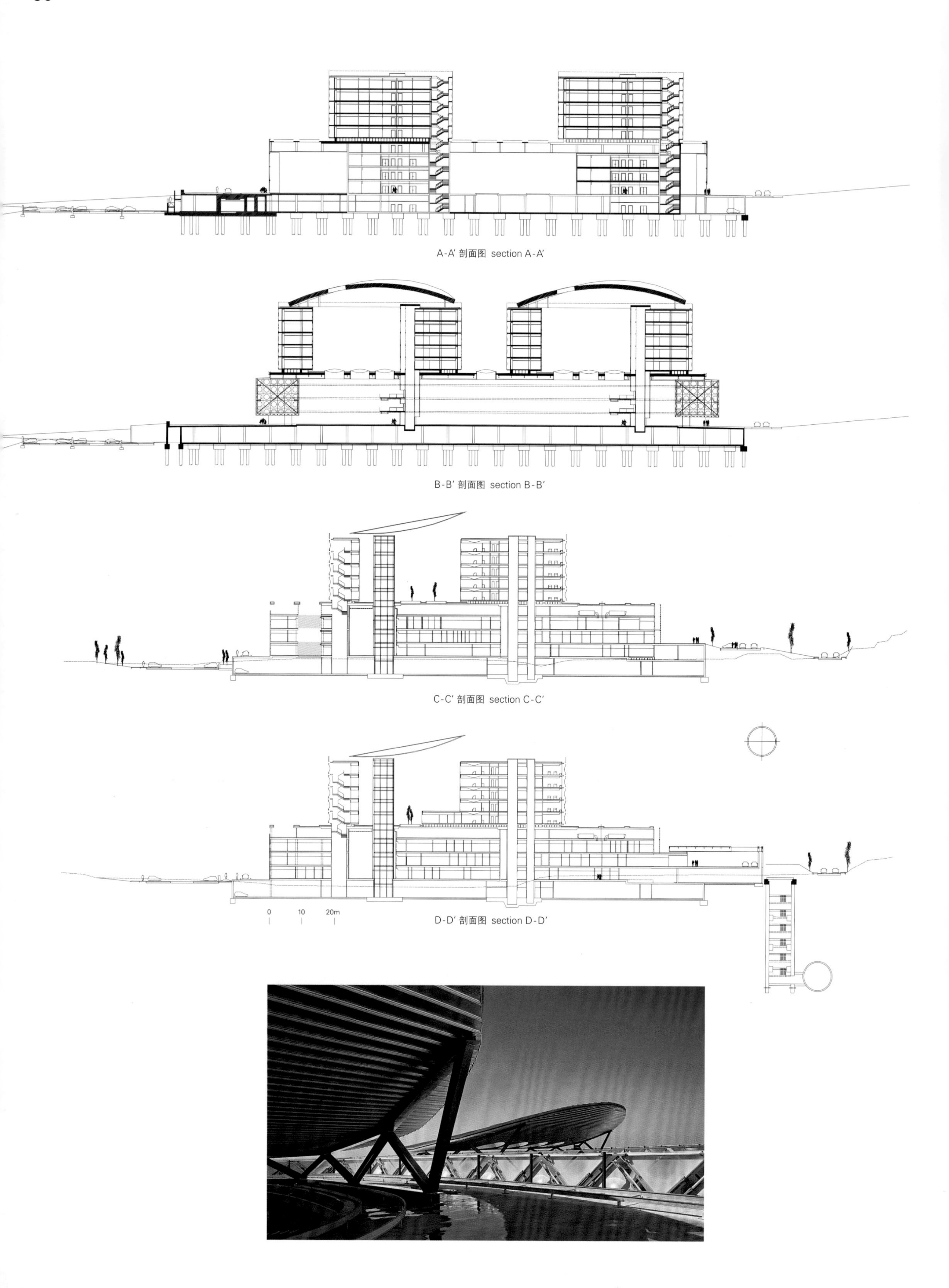

A-A′ 剖面图 section A-A′

B-B′ 剖面图 section B-B′

C-C′ 剖面图 section C-C′

D-D′ 剖面图 section D-D′

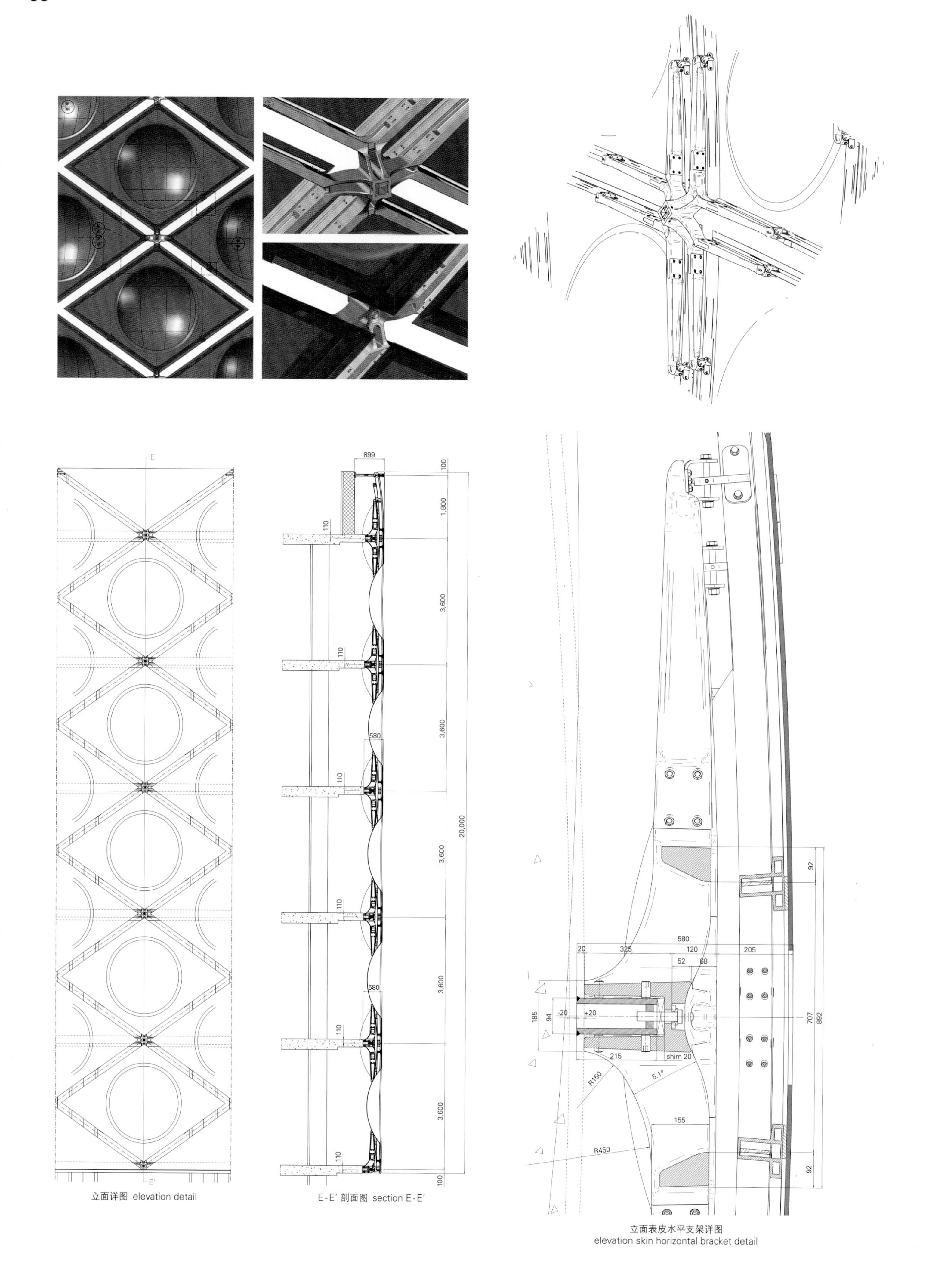

立面详图 elevation detail

E-E' 剖面图 section E-E'

立面表皮水平支架详图
elevation skin horizontal bracket detail

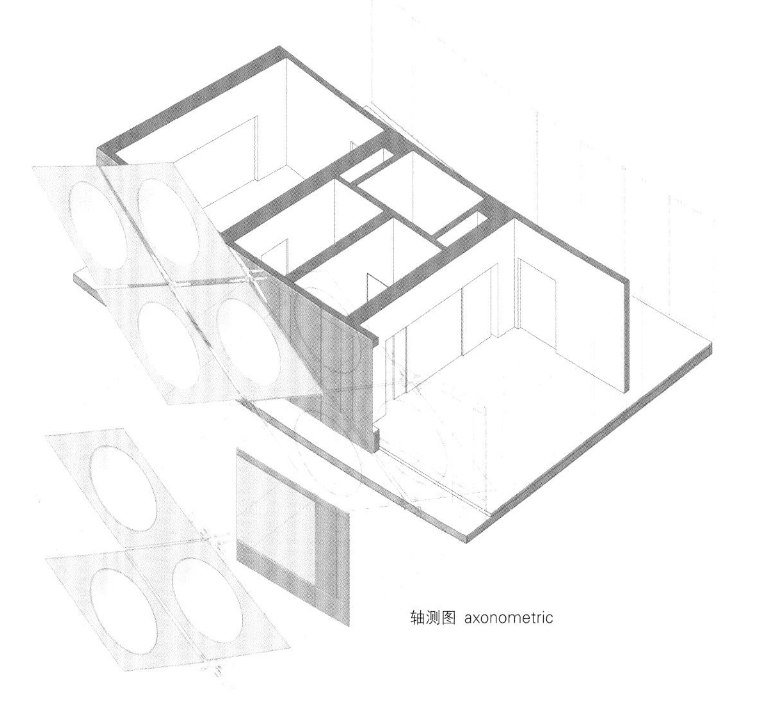

轴测图 axonometric

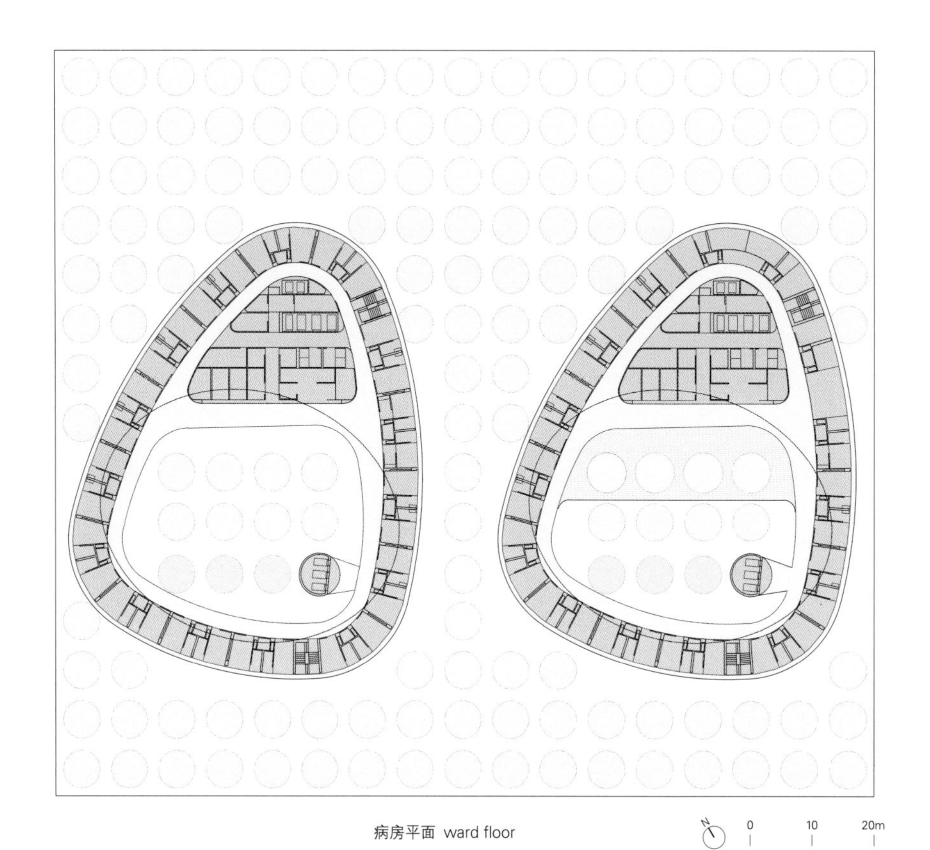

病房平面 ward floor

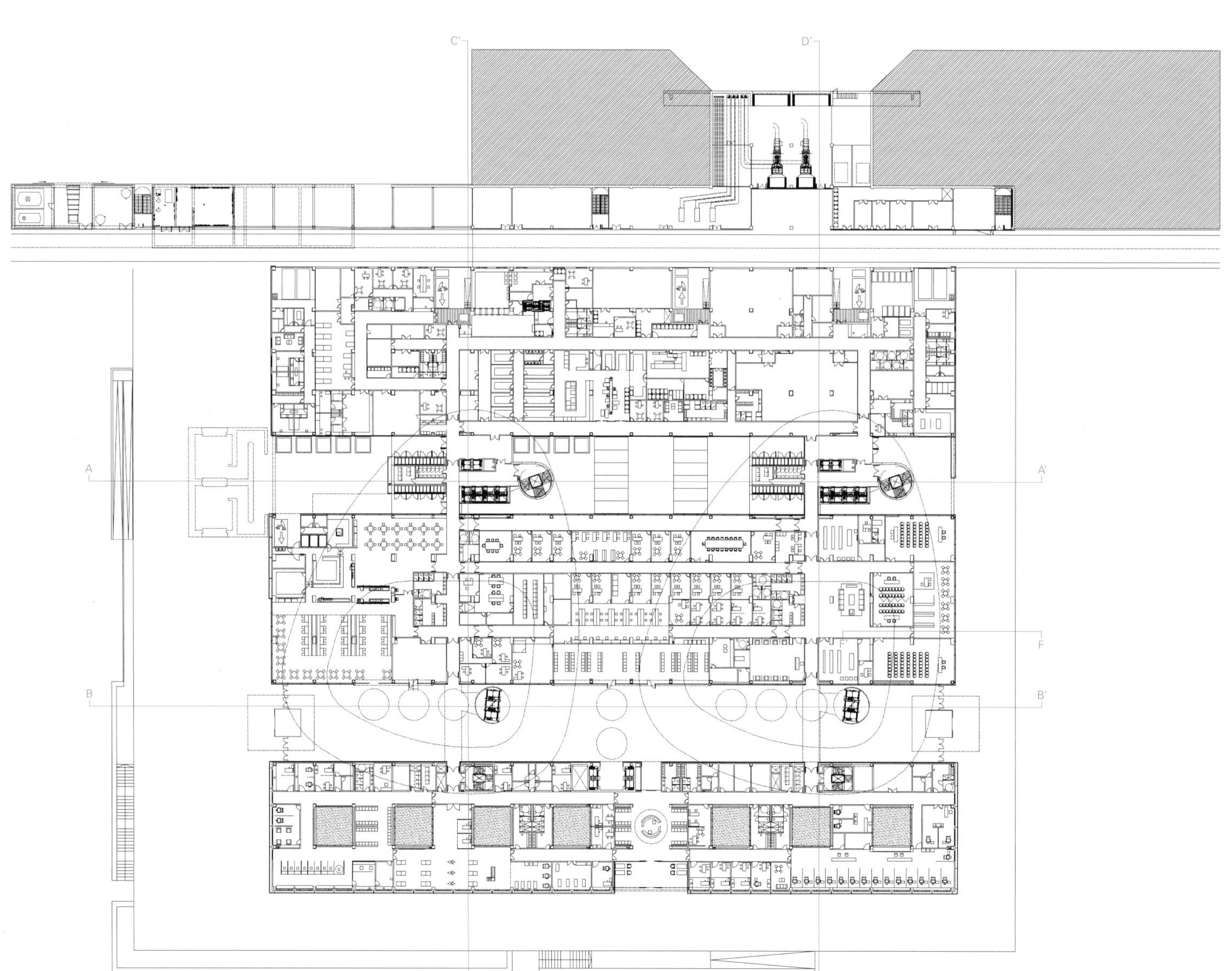

地下一层 first floor below ground

Achieving a perfect functional relationship between the areas, it also provides maximum flexibility to the scheme between the exterior and interior spaces. Differentiating between internal and external circulation, it distinguishes the accesses and the specificity of vertical communication cores, allowing clarifying the scheme.
The position of the two towers, responds to the functional need to have an immediate access to the operating rooms, delivery rooms, emergency and diagnostic. The functionality of the building lays in the way of life the patients and their relatives have during the time they have to be inside.
The strategy is the organization between all the elements that are involved in the project. The three prisms of the basement organize the circulation in an exemplary manner. In the two prisms of the edges are arranged for one to be used externally and the other internally; the block in the middle shares uses when internal and external circulations are necessary, and always without interference.
One of our main points in the building is the sustainability, considering the conditions of solar orientation, topography, built environment and the greenery nearby, without forgetting the urban conditions of application.
The architects incorporate in the system green materials and renewable energy technology, with the objective to save resources and optimize operating costs, providing through the green roof the natural light and ventilation to the inside of the building.
Rafael de La-Hoz Arquitectos

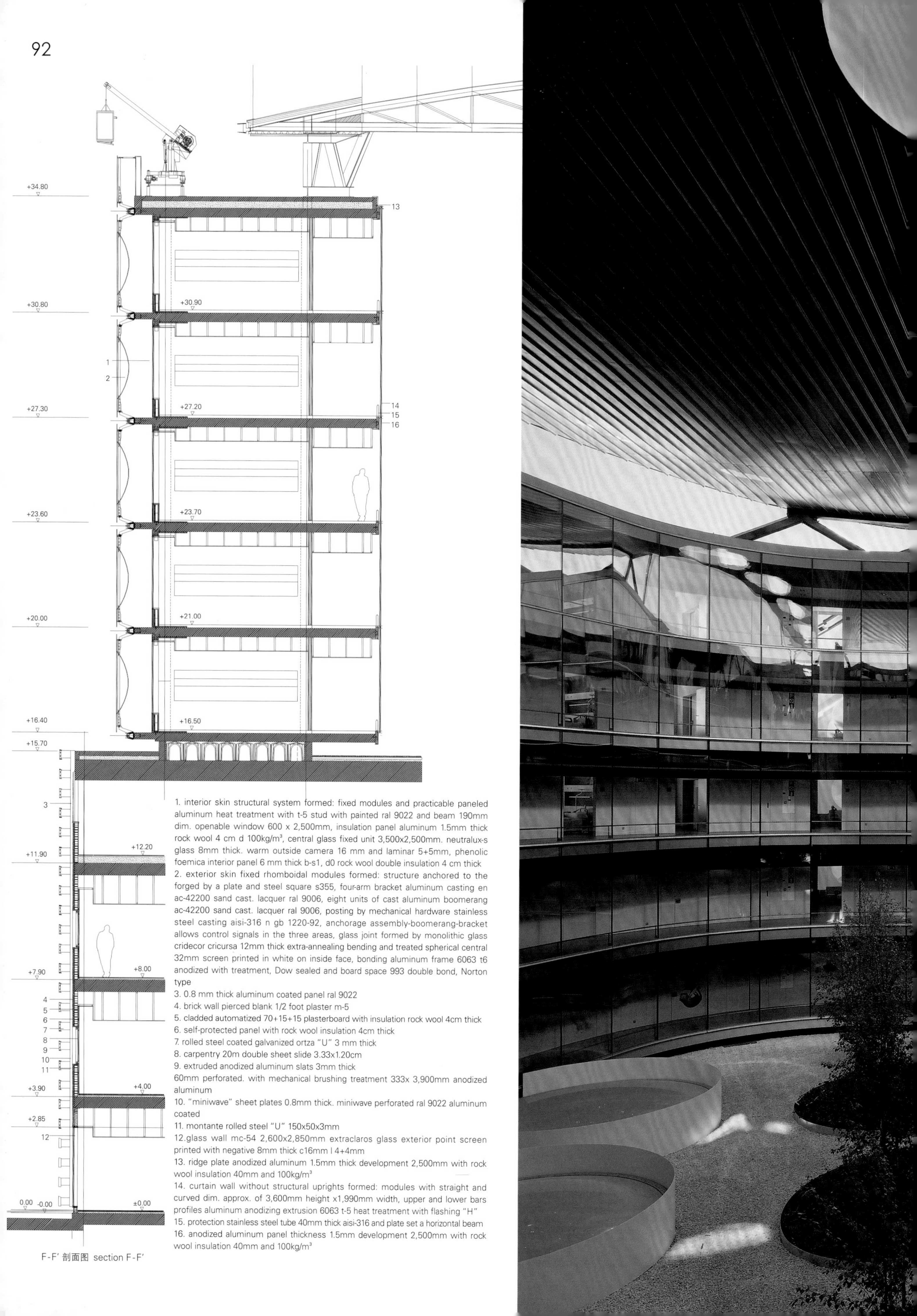

1. interior skin structural system formed: fixed modules and practicable paneled aluminum heat treatment with t-5 stud with painted ral 9022 and beam 190mm dim. openable window 600 x 2,500mm, insulation panel aluminum 1.5mm thick rock wool 4 cm d $100kg/m^3$, central glass fixed unit 3,500x2,500mm. neutralux-s glass 8mm thick. warm outside camera 16 mm and laminar 5+5mm, phenolic foemica interior panel 6 mm thick b-s1, d0 rock wool double insulation 4 cm thick
2. exterior skin fixed rhomboidal modules formed: structure anchored to the forged by a plate and steel square s355, four-arm bracket aluminum casting en ac-42200 sand cast. lacquer ral 9006, eight units of cast aluminum boomerang ac-42200 sand cast. lacquer ral 9006, posting by mechanical hardware stainless steel casting aisi-316 n gb 1220-92, anchorage assembly-boomerang-bracket allows control signals in the three areas, glass joint formed by monolithic glass cridecor cricursa 12mm thick extra-annealing bending and treated spherical central 32mm screen printed in white on inside face, bonding aluminum frame 6063 t6 anodized with treatment, Dow sealed and board space 993 double bond, Norton type
3. 0.8 mm thick aluminum coated panel ral 9022
4. brick wall pierced blank 1/2 foot plaster m-5
5. cladded automatized 70+15+15 plasterboard with insulation rock wool 4cm thick
6. self-protected panel with rock wool insulation 4cm thick
7. rolled steel coated galvanized ortza "U" 3 mm thick
8. carpentry 20m double sheet slide 3.33x1.20cm
9. extruded anodized aluminum slats 3mm thick
60mm perforated. with mechanical brushing treatment 333x 3,900mm anodized aluminum
10. "miniwave" sheet plates 0.8mm thick. miniwave perforated ral 9022 aluminum coated
11. montante rolled steel "U" 150x50x3mm
12.glass wall mc-54 2,600x2,850mm extraclaros glass exterior point screen printed with negative 8mm thick c16mm I 4+4mm
13. ridge plate anodized aluminum 1.5mm thick development 2,500mm with rock wool insulation 40mm and $100kg/m^3$
14. curtain wall without structural uprights formed: modules with straight and curved dim. approx. of 3,600mm height x1,990mm width, upper and lower bars profiles aluminum anodizing extrusion 6063 t-5 heat treatment with flashing "H"
15. protection stainless steel tube 40mm thick aisi-316 and plate set a horizontal beam
16. anodized aluminum panel thickness 1.5mm development 2,500mm with rock wool insulation 40mm and $100kg/m^3$

F-F' 剖面图 section F-F'

公共安全
Public Safety

在现代城市元素中，一些建筑可归为公共安全范畴，体现了有趣的概念模糊性。

虽然公众代表城市建筑的受益者，但在定义建筑的职能时，他们常发挥次要作用，因为通常游客仅允许参观少数特定地点。

通过分析遍布欧洲的七个工程，可发现该主题比看起来复杂，这些建筑和当地居民的关系确实仍然存在，并且关系主要集中在外部，尤其是这些建筑特别吸引参观者，通过有力的外形来影响周边。如果我们考虑城市区域的景观构成和建筑类型，会发现这种建筑与场地之间的关系与18世纪人们所经历的风景如画的自然环境和公共新古典建筑是非常相似的。根据这一设计思路，借助旧建筑模型，使建筑作为"城市"结构融入到景观中，再采用全新的形状和材质来重获新生。

同时，在这些迷人的外围护结构内，常隐藏着前沿的室内设计和空间，从而创造出现代的工作场所。

Among the components of the contemporary city, the buildings that can be classified under the public safety category embody an interesting conceptual ambiguity.
Although the community represents the beneficiary of their construction, the general public often plays a secondary role into the definition of their brief, because visitors are usually allowed in few specific areas only.
Through the analysis of seven projects across Europe, it can be shown how this subject is more complex than it appears, indeed, the relationship between these constructions and the local population still exists and it is mainly external. In particular, these buildings engage the viewers thanks to the images of the powerful shapes they project towards the surroundings. If we consider this scenic composition of the urban area and the type of the constructions, the connection of object-address is very similar to that one experienced by people in the picturesque sequences of nature and civic Neo-Classical buildings during the 18th century. Following this perspective, their presence into the landscape as "civic" structures refers to old models, renewed by fresh shapes and materials.
At the same time, into these fascinating envelopes is frequently hidden a cutting edge interior design and spaces that create a contemporary workplace.

新城市风格
Neo Civic

公共建筑的概念自城市概念化以来便开始演变，但有些建筑类型显示出了异于其他建筑的演变进程：即用于公共安全的建筑。

民用建筑的演变既包括建筑的实质层面，也包括社会看待它们的方式。在这一过程中，有一些功能更能够描述和象征一个时代或一种社会的精神。例如，刘易斯·芒福德以博物馆为范例描述了一个转型中的社会，尤其指出该机构最初作为文化项目与在20世纪被商业化以后的差异。从某种角度来看，民主化和资本化已经重塑了诞生于18世纪的博物馆机构，使其适应新的用户：不再是法国大革命的普通公民，而是有自己的兴趣和背景的现代个体客户。同时，所有的公共空间都拥有相似的趋势，注重开放和互动性。除极少数现象以外，公共安全建筑似乎没有经历这些转变。

现代公用建筑的起源与社会的整体化发展有着密不可分的联系，可追溯到法国启蒙主义时期，为了答复替代了教会和贵族赞助者的政府当局和私有客户，很多建筑类型第一次被定义。此次讨论主题最为重要的一个方面就是这些城市建筑的功能性关联和这些年在景观格局中的存在感。

当今，城市规划能够解决区域覆盖这些方面的问题，这是一种既可从功能性也可从象征性来解读的特性。为了了解某区域是否需要公共安全建筑，相同用途建筑间的距离可作为基本因素。然而，如果我们考虑个人基本安全需要，也有看法认为警察局分布集中与安全观念的联系更为紧密，而不是立即实施求助或抓捕行动的可能性。该因素的重要性不是在整个城市平均分布的，实际上与城市化地区的边缘更为相关。这些居住在刚刚变为城市化边界的人们，由于不了解，尚未建立起对该地区的信心。在塞维利亚，向东扩展的区域完全符合这一描述。由Paredes Pedrosa建筑事务所设计的警察局却是例外，它打断了高层居民楼整齐划一的布局，形成了几何城市构造。据建筑师介绍，设施的辐射状布局受

The concept of public architecture is something that has evolved since the conceptualisation of the city, but there are types of buildings that have shown a different evolution from the others: those for public safety.
The evolution of civic buildings has involved both the physical sphere of the architecture, as well as the way in which the community looks at them. During this process, some functions more than others have been able to describe and embody the spirit of an era or a type of society. For example, Lewis Mumford described a shifting society using the museum as paradigm, in particular pointing out the difference between the original scope of this institution as a cultural enterprise and its commercialization during the 20th century. From a certain perspective democratization and capitalism have reshaped the 18th century born institution of the museum adapting it to the new users: not the generic citizens from the French Revolution but the contemporary monadic clients with their own interests and backgrounds. At the same time the whole public space shared a similar trend, with openness and interaction as leading words. Excluding minor phenomena, this does not seem to have happened for the public safety buildings.
The origin of the contemporary civic architecture, with its connection to the totality of the community, can be traced back to the French enlightenment period, in which several types of buildings were defined for the first time, in order to give an answer to the new aims of public authorities and private clients, which were replacing ecclesial and noble patrons. The most important aspect of this subject for today's themes of discussion is the functional relevance of those buildings for the city and the sense of presence they had in those years within the landscapes.
Nowadays city planning deals with these aspects in terms of area coverage, which is a parameter that can be interpreted both in a functional and in symbolic way. In order to understand whether an area requires a safety public building, the distance among structures with the same purpose can be pointed out as the fundamental factor. However, if we take into consideration the basic personal need of safety, it can be argued that the closeness among police stations relates more with the perception of security, rather than the possibility of immediate intervention. The importance of this factor does not have an isomorphic distribution across the whole city, in fact it is more relevant at the margins of the urbanized area. It is there where the sense of confidence, which belongs to the knowledge of the places, has not yet to be developed by the people that occupy these recently urbanized outskirts. In Seville the new area towards East fits perfectly such a description. There, the *Police Station in Seville* by Paredes Pedrosa Arquitectos is the exception that breaks the monotony of the high-rise dwell-

照片提供：X-TU Architects (©JM Monthiers)

圣丹尼斯警察局作为这座城市的标志性大楼矗立在街道拐角
Saint-Denis Police, which is located at a street angle, will stand as an emblematic tower of this city

到了圆形监狱的启发：这是在18世纪末由哲学家杰里米·边沁提出的结构理论，他尝试在只有一名狱警的情况下，实现同时监管若干名囚犯的目的。将这一方案变换为四座翼楼赋予警察局很高的象征意义，而且与在视觉上具有冲击力的建筑外形相符。设计师结合了三层：地面折叠式设计，将建筑与地面牢固地连接起来，以及两层重叠的折叠带，一层为混凝土，另一层为玻璃（主体结构），构成建筑顶部的特征。混凝土的粗糙质感通过清晰的雕塑轮廓得以加强，散发着与建筑功能相匹配的庄严之感，然而该设施由于高度与规模都很适中，在周边环境中并没有显得很突出。

与这种动态和谐感相似的是位于法国圣丹尼斯由X-TU（Anouk Legendre & Nicolas Desmazieres）完成的警察局项目。在巴黎犯罪率居高不下的区域，该建筑除了起到了警察局的实际功能之外，还能让居民不断增强对政府存在感的认知。非常有趣的是，除国家宪兵团外，1791年法国最初建立警察局也是本着相同的使命。因此，除了有简单而令人愉悦的抽象环境，警察局公共大厅的内部设计还获得了更为深层的象征意义：印有1789年法国大革命的口号（自由、平等、博爱）[1]，并且辅以动作图案，仿佛该口号在全国范围内流传。再仔细一看，公共大厅入口的复杂设施清晰地将内部和外部分开，内外之间主要通过丝网印花玻璃分隔。空间内设计了明确的围护结构主要是出于警察局对于入口安全的要求，允许普通人进入的区域是很有限的一部分。另一方面，在建筑的雕塑外形与定义了巴黎郊区这一区域的实际限制条件、结构和一致性之间创造联系，从而与居民建立象征性关系。

相同的设计策略也被Lamazeta建筑事务所用于设计西班牙曼萨纳雷斯警察局，重新利用老酒庄结构，是在建造博物馆的过程中设计发生变化而来的。从视觉角度来看，法国和西班牙这两个警察局都为雕塑设计，将建筑实体当成主要的交流特色。玻璃的运用丰富了巴黎建筑的外

ing buildings placed neatly to form a geometric urban composition. According to the architects, the facility radial plan has been (freely) inspired by the Panopticon: the structure theorized at the end of the 18th century by the philosopher Jeremy Bentham, who was trying to achieve the best solution for overseeing several inmates with a single observer only. The transposition of that scheme into the four wings that form the building has produced a police station with a high symbolic value, to which corresponds a visually powerful shape. The designers combined three layers: the folded plane of the ground, which deeply connects the building to the site, and two overlapped folded bands, one in concrete and the other in glass (for its major part) that characterize the top part of the structure. The brutality of the concrete, strengthened by the clear sculptural profile, emanates a sobriety perfectly in tune with the function of the building, nevertheless the facility does not dominate its surroundings thanks to its horizontality and appropriate scale.

A similar kind of dynamic harmony is the result of the project by X-TU Architects for the *Saint-Denis Police Station*. In a high crime rate district of Paris, in addition to its actual function as police station, the building also serves to increase the consciousness of the citizens about the presence of the government. It is interesting to note that this was the same mission of the commissariats in France when they were established in 1791 in addition to the Gendarmerie corp. Therefore the interior design of the commissariat public lobby acquires a broader symbolic meaning than being a simple delightful abstract environment: prints of the 1789 French Revolution motto (liberté, egualité, fraternité)[1] are patterned with adverbs of motion as if they were an expression of its spreading across the country. Taking a closer look, the complex configuration of the entry to the public lobby reveals the clear separation between interior and the exterior, which communicate mainly through silk screen printed glasses. The unambiguous enclosure levels within the spaces are primarily due to the access security requirements of a police station, which allow the common people to be admitted to very few parts. On the other hand, a metaphoric relationship with the inhabitants has been established by creating a connection between the building sculptural articulation and the physical constraints, textures and alignments that define this part of the Parisian outskirts.

The same strategic approach has been pursued by the Estudio Lamazeta for the *Manzanares Police Station* in Spain, the result of a design variation during the construction of a museum, which was going to re-use a structure of an old winery. From a visual point

照片提供：ACXT Architects(©Adrià Goula)

西班牙雷乌斯的112急救大楼是一种全新的建筑类型，将加泰罗尼亚地区所有的急救实体都集中在一起
112 Emergency Building in Reus, a new architectonic typology which brings all the bodies of emergencies in Catalonia

照片提供：Katušic Kocbek Arhitekti/Produkcija 004 (©Miljenko Bernfest)

萨格勒布紧急服务中心以白色拉伸材料包裹着简洁的钢立方体
Zagreb Emergency Terminal, the pure form of steel cube enfolded by white tensile materials

观，而另一座建筑中铺砖的形式使得建筑从过分简化变为简洁精炼的构造。博物馆最初在设计立面时，没有打算留出开口，这一观点得到了认可。灯光和空气通过对角线切口从外面进入办公室，切口从外层一直延伸到接待处所在的中心区域。警察局的白色体量保留着以前酒庄以及位于建筑前面的老式结构的特征：仅有若干小窗户和一个宽阔入口的实心墙，这是与外界交流的唯一通道。警察局和道路对面的建筑形成的二元系统在曼萨纳雷斯郊区的这一区域如同一座环境舒适的小岛，这里居民区和工业区相交，开阔空间被基础设施所隔断。这种重叠设计的结果就是典型的城市郊区景观，形式上并不稳定，并且人们一般都是在车内领略其风光。正因为这样，建筑的清晰线条才更便于移动中的人察觉。

新的公共安全建筑通常远离市中心，坐落于城市边缘，很明显是因为需要服务于最新开发项目，但同时也因为只有在这些地方，不断扩大的现代紧急设施才能够找到自己的位置。一座综合设施在特定区域内不断发展变化主要得益于对紧急情况采用的综合方法，与过去相比，它越来越多地将不同部门（警察、消防队、民事和卫生防御）联系成一个整体。确实，在将112定为所有求助请求的电话号码的国际标准出台之后，各分支的混乱状况得到改善，表明地理位置的统一能提高各部门的效率，控制操作成本。由ACXT建筑事务所设计的西班牙雷乌斯112急救大楼便是如此，所有不同的操作空间都聚集围绕在振奋人心的三层高的空间之中。与社区的互动，即积极的服务再一次成为该项目的中心，通过这一理念，人们可了解建筑与基础设施和周围景观的关系。建筑不同部分的构成令人回想起新古典时期建筑，从项目的早期草图与乔万尼·巴蒂斯塔·皮拉内西（1720—1778年）所画的"精致的新古典建筑门廊和立面"之间的对比便可看出一二：与地面的关系通过地下室来实现，在地下室上面为公共区域，再上面是两层功能楼层，被边廊式倾斜构件所包围。建筑的纯白色在晚上就会消失，因为灯光会穿透建筑立面的塑料

of view the two police stations, the French and the Spanish, are both sculptural objects, in which solid masses have been used as the main communicative feature. The use of the glass enriches the facade of the building in Paris, while in the other a tile's pattern makes the building shift from a simplistic to a simple and refined composition. The initial idea for the museum of having facades with almost no openings has been respected. Light and air come into the offices through diagonal cuts have been carved from the exterior perimeter to the central space, where general reception is. The white volume of the police station keeps alive the features of the former winery, as well as those of the old structure in front of it: solid walls only interrupted by small windows and a singular wide entry, the only point of communication with the exterior. The binary system formed by the police station and the building across the road arises as an island of pleasantness in this part of Manzanares' periphery, where the residential zone meets the industrial district and where open spaces are cut by the infrastructures. The consequence of this overlapping is a typical marginal landscape, unstable in its forms and usually experienced by car. For that reason the clear lines of the building facilitate its perception by the viewer that is in motion.

The new public safety buildings are usually located out of the urban centre or at its margins, obviously because they need to serve the most recent developments, but also because it is in those areas that the increased sizes of the contemporary emergency facilities can find place. The growth in required space for one complex is due to the integrated approach towards the emergency, which, more often than in the past, associates different departments (police, fire brigades, civil and health defence) into a common facility. Indeed, after the creation of the dialling number 112 as an international standard for every kind of help requests, new models of dislocation of the branches have demonstrated how the physical association of them would increase their efficiency and limit the operational costs. This is the case of the *112 Emergency Building in Reus*, Spain, designed by ACXT Architects, where the different operational spaces are grouped all together around an inspiring triple height space. Once again the communication to the community that the service is active is central to the project, and, through this concept, it is possible to understand the relationship of the building with the infrastructure and the landscape. The building recalls a Neo-Classical architecture in its composition of parts, as shown by a comparison between one of the early sketches of the project and "The Portico and Facade of an Elaborate Neo-Classical Building" drawing by Giovanni Battista Piranesi

位于英国巴斯的皇家新月楼是一条由30座住宅组成的长排住宅群，由小约翰·伍德设计，1767–1774年
Royal Crescent in Bath, England, a residential road of 30 houses laid out by John Wood the Younger, 1767~1774

卑尔根主要消防站大楼的曲线外形是沿着场地的外沿弧度而设计的
Bergen Main Fire Station with curved shape of the building following the outer edge of the site

网，塑料网穿插在钢结构之间。建筑从周围树丛中冒出头来，即使其轮廓与水平的景观互相协调，也凭借其独特的外形而富有象征意义。

另一座外形高大、在周围环境中具有主导地位的建筑是由Katušic Kocbek建筑师事务所/Produkcija 004设计的萨格勒布紧急服务中心，不过与雷乌斯建筑相比，这一稳固的建筑使得周围景观更为城市化。这座克罗地亚建筑的魅力源于其纯粹的形式：由白色拉伸材料包裹而成的钢立方体。夜晚当外围护结构被灯光照亮时，七层楼（在地面上）在光柱的照耀下显现，光柱沿着建筑外部表层排列。在内部，玻璃的透明性将所有不同功能集中起来。穿孔水平板和半透明垂直隔墙的基础结构使得整个建筑看起来可穿透，并且主要的部分是外露的。从建筑外看，立面上不同的阴影描述了一座超过六层高的建筑的复杂构成，其中包括医院、诊所、管理和教育区域，以及容纳170辆救护车的车库。因此，这一急救站体现了高度复杂功能性与最简洁形式的双重性。

以上所描述的建筑认识到其在社区所发挥的作用，将其庄严的外形与周围环境清晰地分离，创造了与周围建筑的关系。这里提到的二分法不是基于简单的对比，因为可能会强调出所感知的建筑、周围环境和蜿蜒的道路是如何产生张力并带来某种和谐感的。这也是设计新古典式花园的指导原则：曲折的小路通常使得景色更为开阔，建筑通常为各个层次中最为重要的元素。为了完全理解这一平行关系，替换一些元素非常重要，在原先的元素之间保持相同的关系：基础设施干线取代了我们现代景观中的慢行道，其他建筑通常为一般的建筑构造，正如树丛通常所发挥的作用——遮蔽了景色，而新公共设施取代了新古典主义建筑，成为该社会引以为荣的标志。位于挪威卑尔根的主要消防站就是这样的标志之一。建筑的地面设计以两个同心曲线为特征，与该景观两种不同的风格对应。建筑师将外部铜板曲线表面与切断峡湾的道路相对立，从而保护了设施并且得到了与周围环境相和谐的规模和表达方式。

(1720-1778): the relationship with the ground is achieved through a basement, on top of which lays the public floor surmounted by two storeys of the operational box, wrapped by a peristyle of slanted structural members. The solid white colour of the building disappears at night time because of the light passing through the plastic meshes of the facades, which are interposed among the steel structure. The building stands over the surrounding carpet of trees gaining a powerful iconic presence in virtue of its alien shape, even though it is related to this horizontal landscape by the articulation of its silhouette.

Another building which has a burly dominance that creates the context is the *Zagreb Emergency Terminal* by Katušic Kocbek Arhitekti/Produkcija 004, although the presence of a robust infrastructure makes the landscape more urban in comparison with the site in Reus. The fascination that the Croatian building inspires derives from its pure form, a steel cube enfolded by a white tensile material. When the envelope is lit at night, seven floors (above the ground) are revealed beneath the columns of the lights, which are placed along the external skin. Internally, it is the transparency of the glass that brings together all the several functions. Reading the basic construction of perforated horizontal slabs and translucent vertical partitions the whole building appears to be permeable and its essentiality is exposed. For an external viewer the different shadows in the facades describe the complexity of the composition in one building of a hospital, a clinic, administrative and educational building spaces with a garage for 170 medical vehicles over six storeys. Therefore this terminal embodies a duality of a highly sophisticated functionality combined with one of the simplest forms.

The architecture so far described stand with a consciousness of the symbolic role they play for their communities, creating relationships with the sites they are in based on a clear separation of their serious shapes and the surroundings. The dichotomy described is not based on a simplistic contrast because it is possible to underline how the building, the surroundings and the winding roads from where they are perceived, generate a tension that brings a certain harmony. This was also a leading principle in the design of Neo-Classical gardens: curved paths used to open sceneries in which a building was usually hierarchically the most important element. To fully understand this parallel it is important to replace some elements, keeping the same relationships among the originals: infrastructural arteries replace the slow paths in our modern landscapes, other buildings, usually architectonically average, work as the trees used to, screening sceneries, while new

照片提供：Bergmeister Wolf Architekten (©Günter Richard Wett)

Margreid消防站大楼是在岩石中挖了三个巨大的洞穴而建成的
Margreid Fire Brigade, three big caverns are drilled into the rock

然而，玻璃立面的风格让人想起了位于巴斯的皇家新月楼[2]，向管辖区内在建筑外的人们展示了消防工作者的活动。建筑的另一个重要元素是防火训练塔，正如设计师所说的，"作为设施中起统一作用的物体，象征了它在社会中的重要地位"。

在新古典主义花园中，不仅是那些大楼，就连废墟和窟穴也是关键元素，因为它们经常在景观中添加一些出其不意的元素，当然也会产生意大利Margreid消防站大楼给参观者带来的惊奇感。Bergmeister Wolf建筑师事务所设计了黑色墙体，背倚南蒂罗尔红酒产地的山脉，作为一面屏风界定了嵌入石头中的实际建筑的范围。虽然该项目拥有极高的美学价值，但这种极简式设计和将消防站嵌入地面中的决定完全是出于节能考虑。另外，通过这种方法，建筑师将珍贵的土地留给了葡萄种植，尽可能降低运营成本，从而显示建筑师对公众利益的重视。该建筑仅有三个很小的体量凸出于墙体向外界显示其存在，而一个很小的红色专用标志也是其存在的唯一优美象征。

每一座列举的建筑都可以用严肃、简洁和具有象征性来形容其外观，然而这些建筑所代表的与社会道德的交流这一共同主线贯穿在每个设计中。本文将这些建筑实例和新古典主义建筑相联系来展示共同的主题和相同的思考方式，并不是想要从历史角度来解读这些现象，而是指出现代公共民用设施作为公共建筑的特殊组成部分，仍然表达了我们社会的基本人类学需求，这也是被整个社会所认可的。

1. liberté，egualité，fraternité，(自由、平等、博爱)成为19世纪法国的国家座右铭。
2. 英国，建于1767至1775年间，由小约翰·伍德设计。

1. liberté, egualité, fraternité, (freedom, equality, brotherhood) became the National motto in the 19th century
2. England, built between the 1767 and 1775, designed by John Wood the younger

public facilities stand in place of Neo-Classical buildings as proud signs for the community. The *Bergen Main Fire Station*, Norway, is one of those signs. The building floor plans are defined by two concentric curves, which correspond to the two different attitudes towards the landscape. The architects oppose an external copper plate curved surface against the road that cuts the fiord, in order to protect the facility and gain scale and language appropriate for the context. However, a glass facade, which recalls that one of the Royal Crescent in Bath[2] for its articulation, shows the activity of the firemen to the viewers outside the precinct. Another important element of the building is the tower for fire training, which, according to the architects, works as a "unifying object in the facility and signals its importance for the community in general".

In Neo-Classical gardens not only the buildings, but also small follies, ruins and grottos were crucial elements, because they used to insert a component of surprise into the landscape, and certainly a sense of wonder is what is conveyed to a visitor by the *Margreid Fire Brigade*, Italy. Bergmeister Wolf Architekten designed a black wall against a mountain in the wine area of the South Tirol as a screen to define the actual facility that has been dug into the rock. Although the project achieves a high aesthetic value, the purpose of this minimalism and the decision to insert the station into the ground have a purely energy saving purpose. Besides, in this way the architects left precious land for the cultivation of grapes, which in addition to achieving the lowest running costs possible shows their intense attention to the community interests. Only three small volumes that pop out from the wall reveal outside the presence of the fire station and a small red logo is the only elegant symbol of its existence.

In every one of these examples, the adjectives of sober, simple and iconic can be applied to their shapes, while the communication of the ethic they embody to the community is the common thread running through each design. A parallel has been drawn between these examples and Neo-Classical architecture, demonstrating common themes and similar ways of thinking. The intention has not been to create a historic reading of the phenomena, rather to point out that contemporary public civic facilities are a special sector of civic buildings that continue to be an expression of the basic anthropological needs of our society and which the whole community can recognize itself. Simone Corda

雷乌斯112应急大厦

ACXT Architects

112
emergències
112
emergències

位于雷乌斯的新112大厦为加泰罗尼亚的新应急管理及服务体系树立了典范，它是该国首座获得LEED认证的公共设施。

这是一种新的建筑类型，将加泰罗尼亚所有负责应急管理的机构都汇聚到了一起。

以前，这些机构分散在该地区的各个地方（112呼叫中心、警察局、消防站、公共医疗和民防机构），使用不同的电话号码，彼此之间不共享技术框架。

将它们汇聚到同一个屋檐下，在此它们可以共享技术与操作程序。将所有线路都替换成唯一的应急号码112会使应急管理变得更加高效、更便于协调。

该综合设施遵循的基本原理是借助一套以明确、精准的项目工具为基础的组织体系，使众口难调的需求标准化，并将重点放在人们认为重要的方面：

—场地规模：在该地区内的位置、与基础设施之间的关系以及人造景观。

—社会与操作人员之间的关系（之前处于隐藏状态）：建筑外观、体现出新的应急管理体系、全天候待命。

—操作机构汇聚到一座建筑里：一座容纳下所有机构的结构，统一使用白色，横向贯穿所有机构。

—操作机构之间的关系：建立一个可以促进协同作用的公共空间（此前他们远程共享信息，如今他们可以直接交流）。

该建筑在水平方向上分为三层：底层服务区、公共区域以及操作楼层。底层服务区（停车场、更衣室、储物间、休息区、建筑设备）变成了一个可以适应场地特殊性（地形和形状）的元素。其屋顶把一片景观归还给周围环境，与位于建筑一层的公共区域（礼堂、新闻中心、餐厅）相衔接。位于橄榄树园上方的这一楼层为人们提供了欣赏整个地区的绝佳视角，并且延伸到上面的楼层：盒状操作区。

操作区的外壳由一个巨大的金属结构和一张塑胶网构成，该网可以达到双重目的：从各个角度防止立面受到太阳光直接辐射；打消人们在立面上设置开口的念头，从而增强该建筑的区域特色。

除了解决操作室的功能需求之外，金属结构还将灵活性融入进来，方便将来的楼层增添配置；它借助自身的坚固性和白色调（就操作机构的建筑外观而言，白色比较中庸），强化了汇聚在新应急管理模型中的操作机构的统一形象。

为了促进操作员之间的配合与协作，操作室呈环状分布在一大片空间内，该空间同样有利于自然漫射的太阳光照射到操作区的内部。

若非如此，垂直方向上的沟通将分别在四个核心区展开：电信塔核心区——在接待访客时同样有用、管理核心区、维护核心区以及工人主通道核心区。

这一功能矩阵（水平分层搭配垂直沟通）与该建筑的结构及概念配置不谋而合：一层（底层服务区）安装了网状地板和盒状金属结构，它构成了操作区域，并由四个混凝土核心区支撑。

为了保护各项操作，该建筑在室内和室外都采取了高标准的安全措施。为了满足每天24小时、每年365天的全天候工作需求，主要的建筑设备（电力、电信、空调）必须留有富余，如果供应中断，该建筑应保证可自主运行至少5天时间。

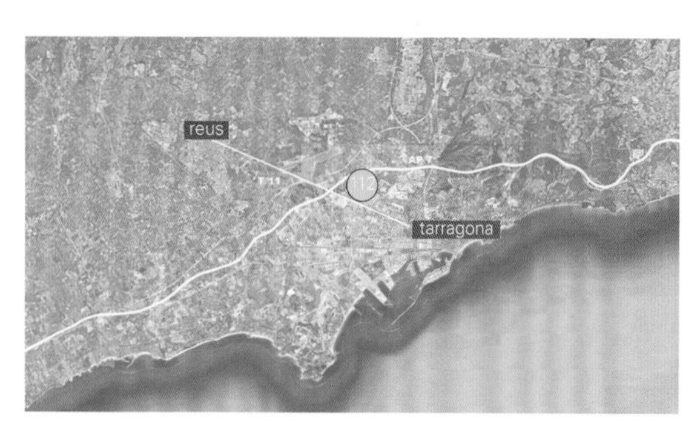

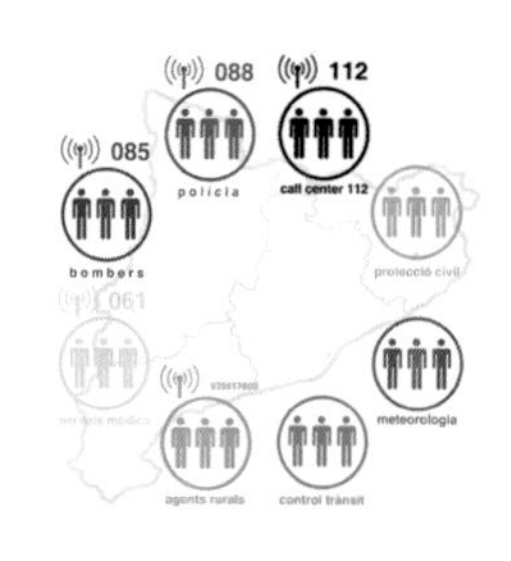

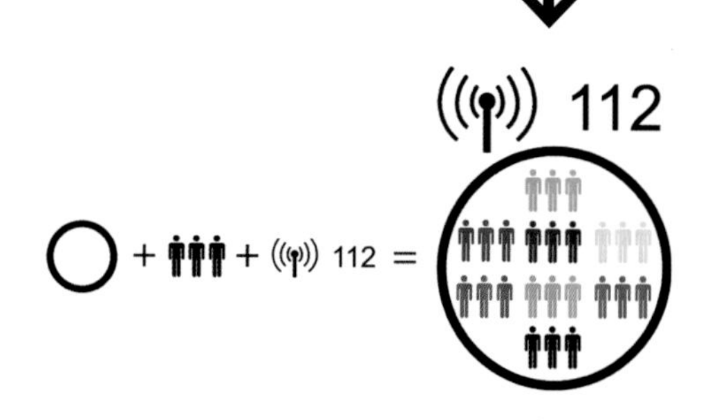

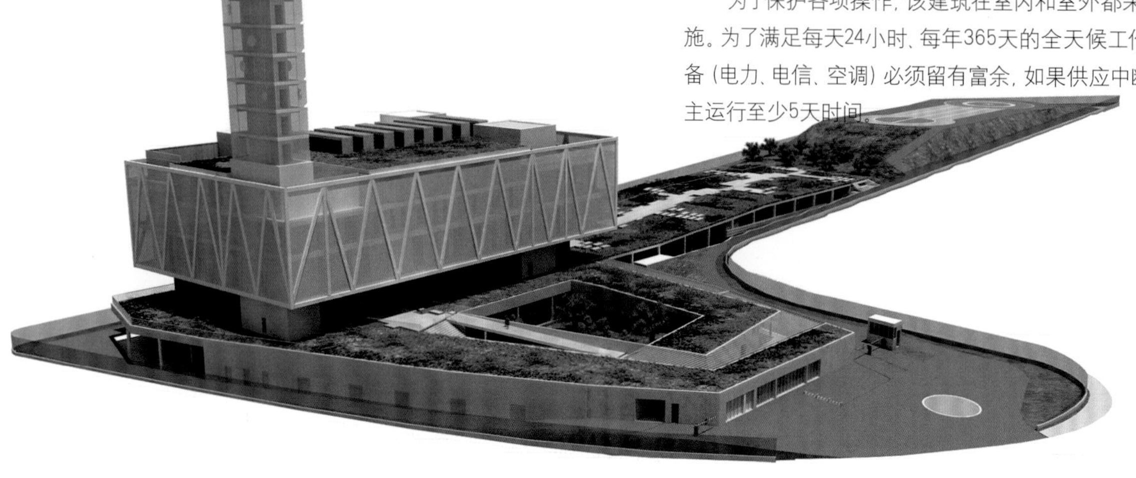

112 Emergency Building in Reus

The new 112 Building in Reus is the model for the new emergencies management and service system in Catalonia, and the first public facility in the country to have an LEED certification.

It is a new architectonic typology that brings together all the bodies in charge of managing emergencies in Catalonia.

Before, these bodies were scattered around the territory (the 112 call centre, the police, firemen, public health and civil defence), had different phone numbers and didn't share technological frameworks.

Their gathering under the same roof, where they share technology and processes, and the substitution of all lines for the single emergencies number 112, will result in a more efficient and better coordinated management of emergencies.

The complex follows the rationale using an organizational system based on clear and concise project tools which help standardize its heterogeneous demands and highlight what we consider important:

- The scale of the place: location within the territory, relation with infrastructures, artificial landscape.
- The relationship (previously hidden) between society and operatives: exterior view of the building, showing the new emergency management system, day and night presence.
- The coming together of the operative bodies in one building: a single structure holds them all, unified by the colour white, transversal to them all.
- The relation between operative bodies: definition of a common space that activates synergies (before they shared information

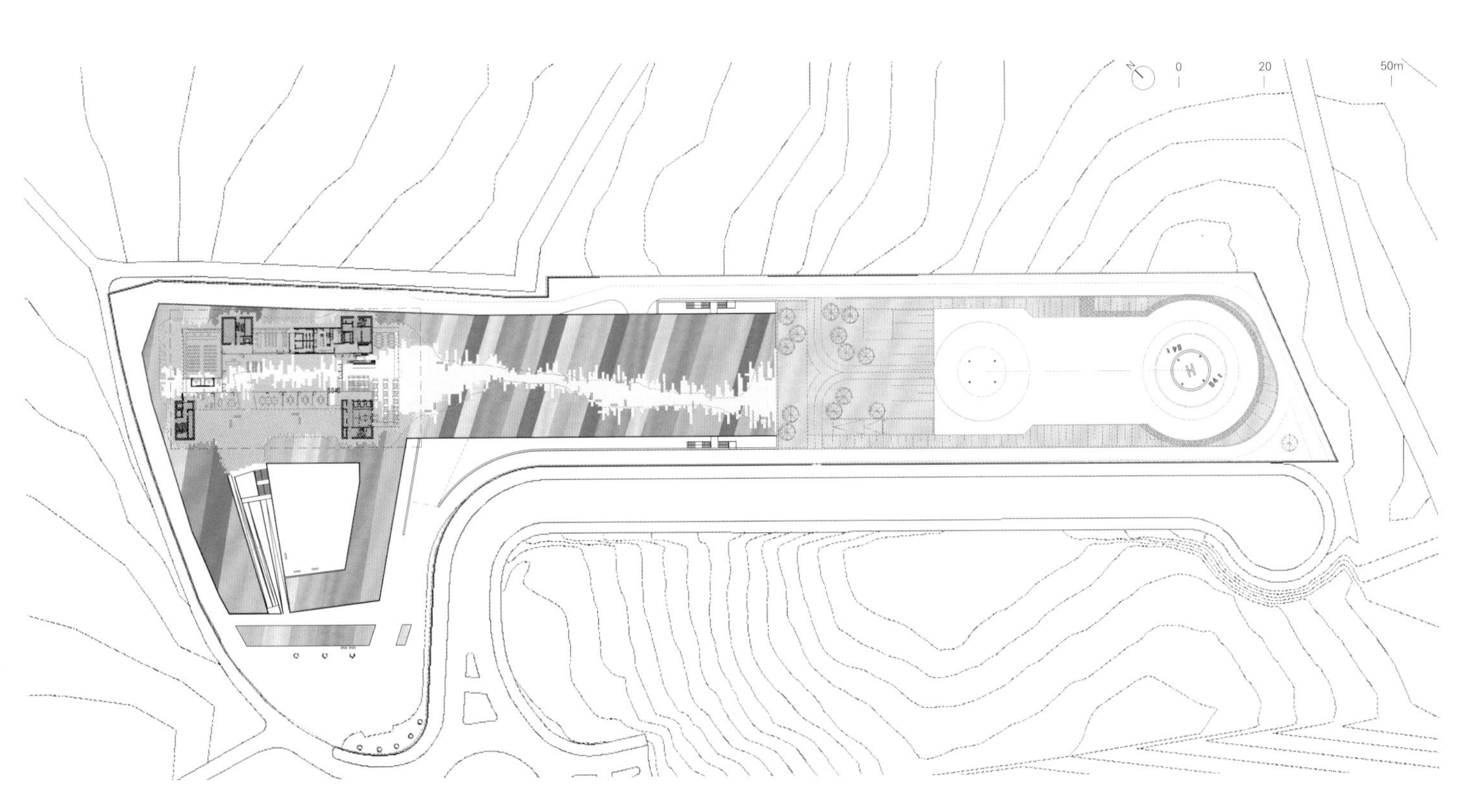

112
emergències

项目名称：112 Emergency Building in Reus
地点：Reus, Spain
建筑师：Marco Suárez Pizarro (ACXT Architects)
项目团队：Élida Mosquera Martínez, Alex Borrás (Bec), Claudia Carrasco, Mireia Adnetller, Sorana Radulescu, Roberto Molinos Esparza
项目管理：Alfredo Fernández Parent
结构工程师：Joel Montoy Albareda, M. del Mar Sahún Argüello, Roger Señís López, Ana Andrade Cetto, Leonardo Domínguez Ferreira
电力工程师：Alex Boada
环境工程师：Pablo Jorge Vispo
照明工程师：Mercedes González Carrascosa
公共卫生设施：Miguel Castro, Pablo Jorge Vispo
电讯：Alfredo Fernández Parent, Vicente Montoya Barrera
可持续性：María Cortés Monforte
总规划：Javier Losada
场地监督：Marco Suárez Pizarro, Carlos Garín Caballero
施工管理：Víctor Amado Valido
总楼面面积：14,985m² 竣工时间：2010
摄影师：©Adriá Goula(courtesy of the architect)

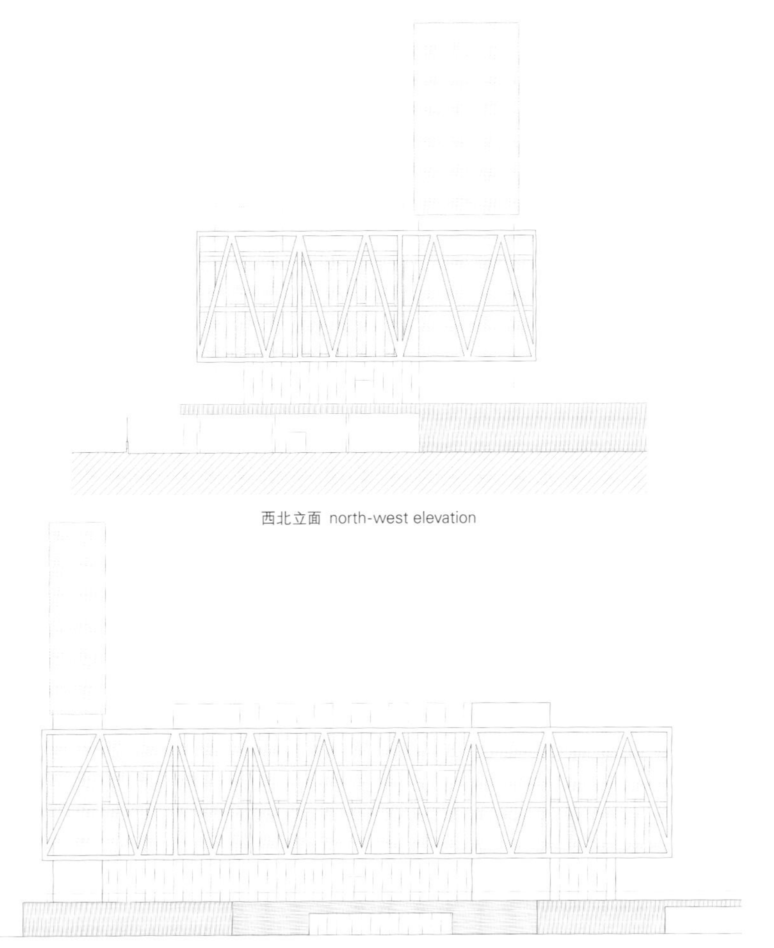

西北立面 north-west elevation

西南立面 south-west elevation

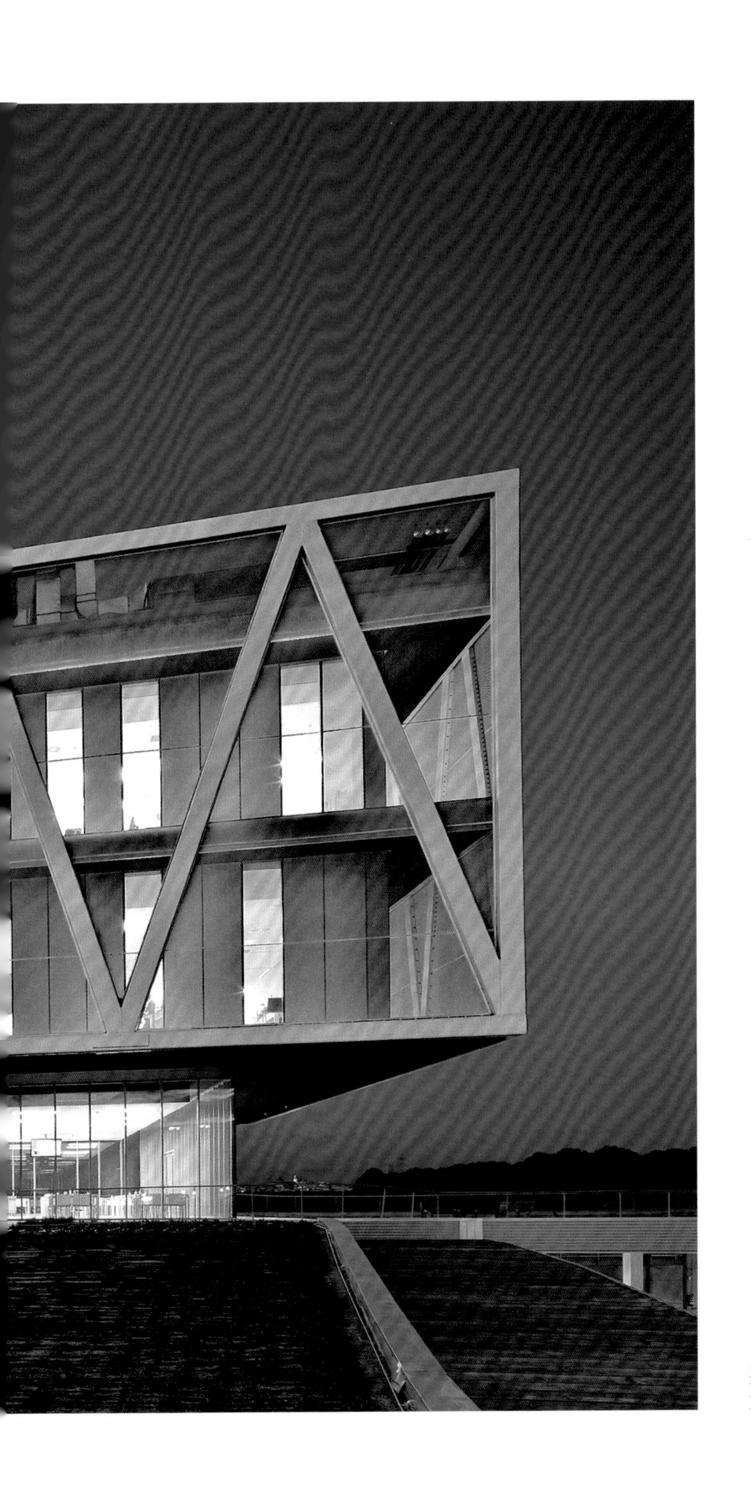

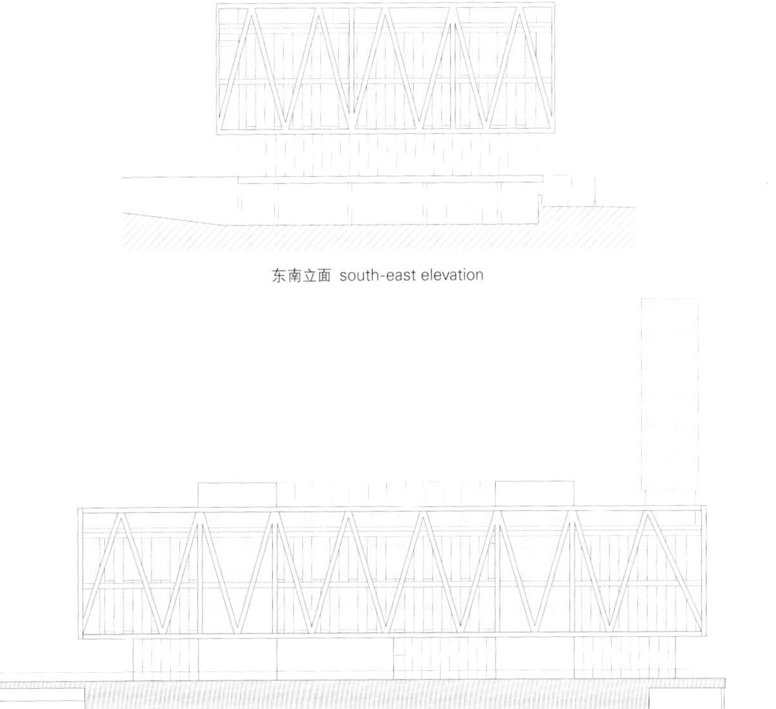

东南立面 south-east elevation

东北立面 north-east elevation

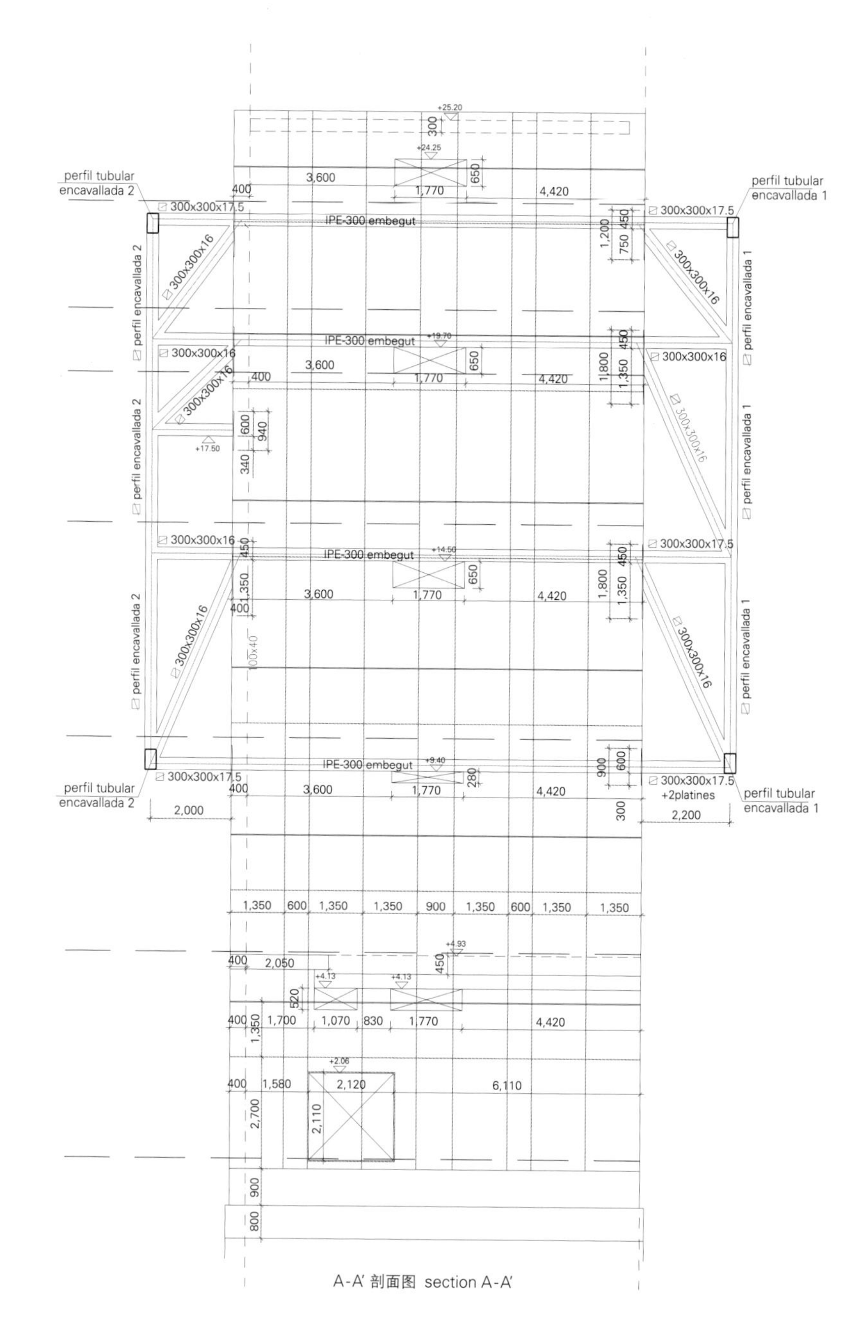

A-A' 剖面图 section A-A'

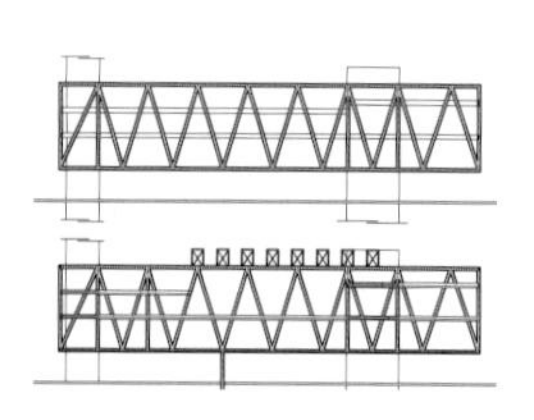

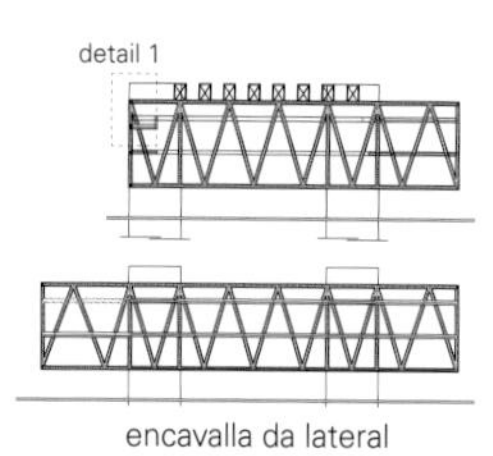

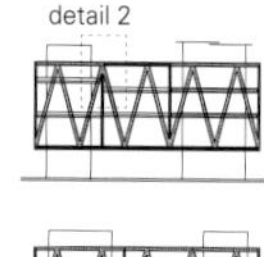

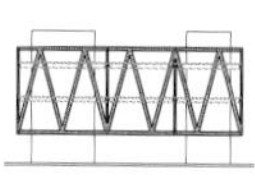

encavalla da lateral

1. 500x300xt
2. soldadura a topail amd preparació d'arestes
3. platina reforç e=20mm
4. platina reforç 1,500x300x16
5. soldadura a topail amd preparació d'arestes (canvi secció)

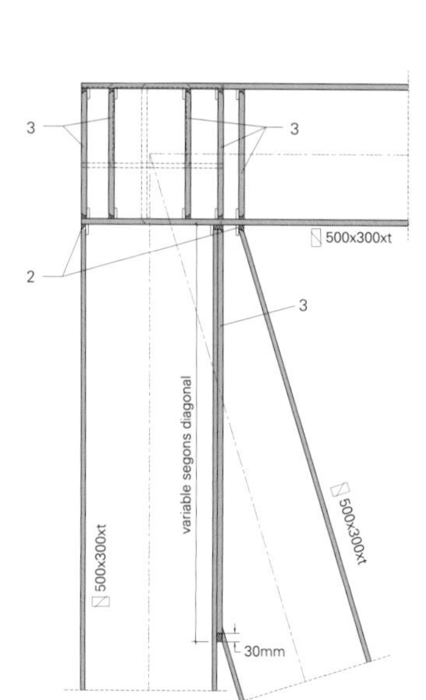

详图1 detail 1

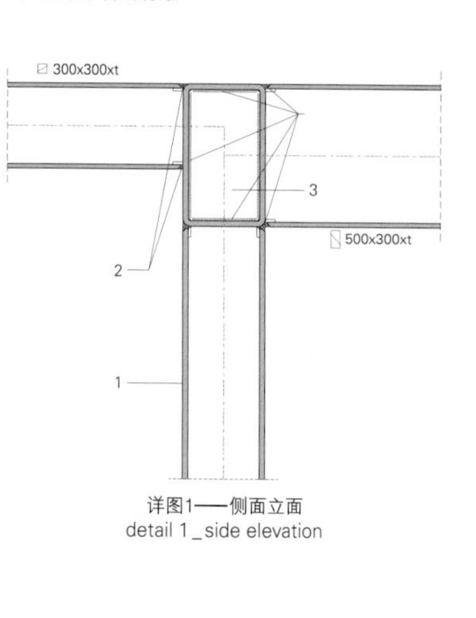

详图1——侧面立面
detail 1 _ side elevation

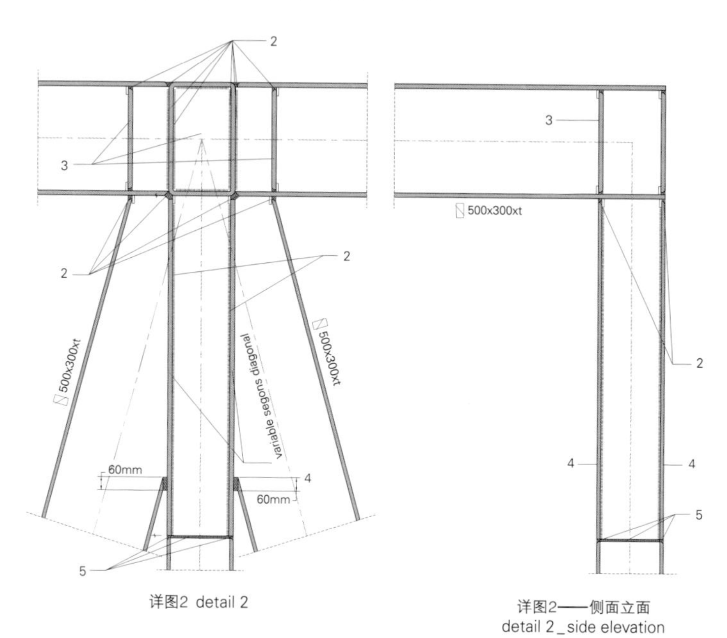

详图2 detail 2

详图2——侧面立面
detail 2 _ side elevation

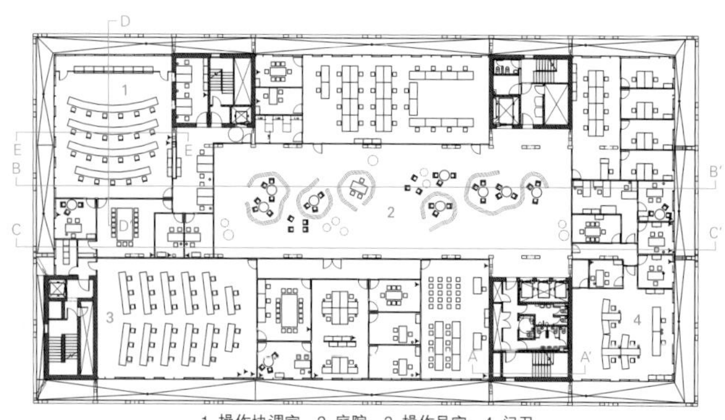

1 操作协调室 2 庭院 3 操作员室 4 门卫
1. operational coordination room 2. courtyard 3. operator's room 4. guard room

一层 first floor

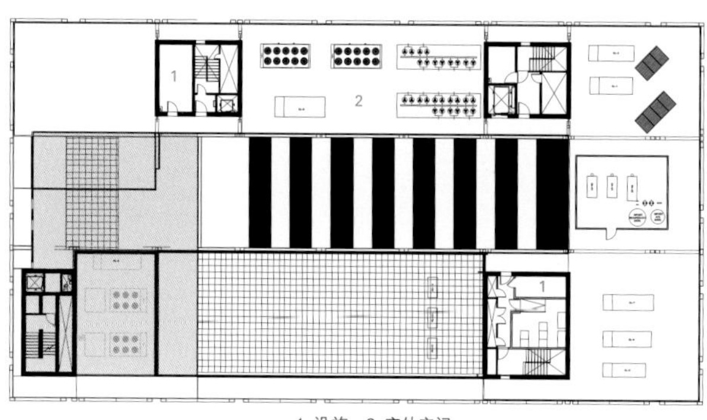

1 设施 2 室外空间
1. facilities 2. outside space

三层 third floor

1 大厅 2 会议室 3 新闻中心 4 主要休息室
1. hall 2. conference room 3. press room 4. main lounge bar

地下一层 first floor below ground

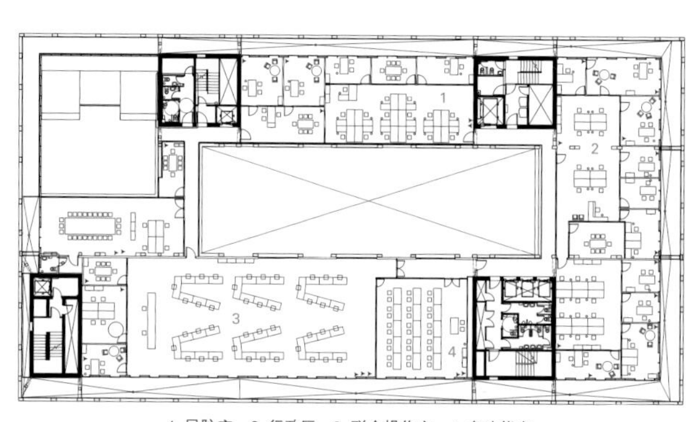

1 民防室 2 行政区 3 联合操作室 4 多功能室
1. civil protection's room 2. administrations 3. joint operations room 4. multipurpose room

二层 second floor

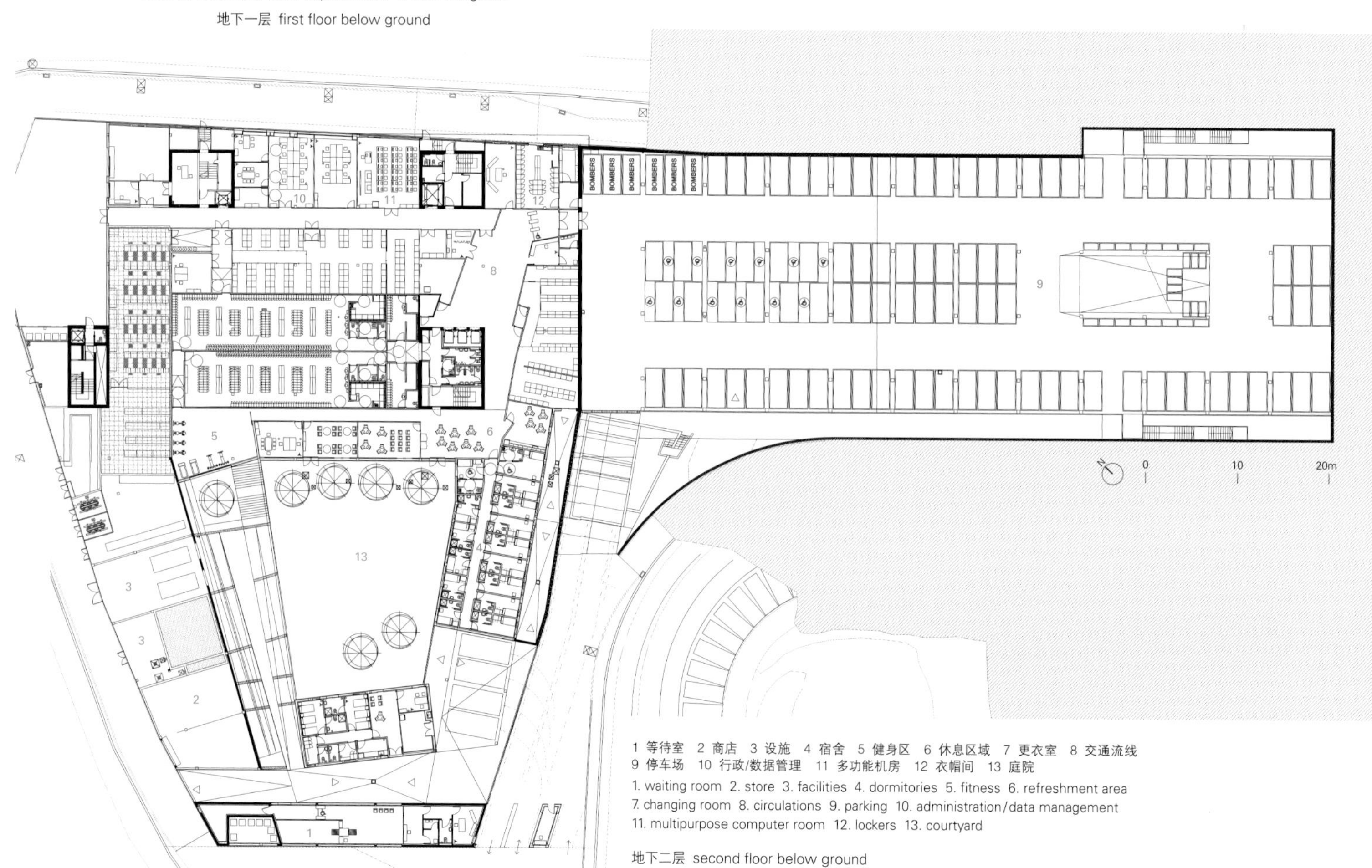

1 等待室 2 商店 3 设施 4 宿舍 5 健身区 6 休息区域 7 更衣室 8 交通流线
9 停车场 10 行政/数据管理 11 多功能机房 12 衣帽间 13 庭院
1. waiting room 2. store 3. facilities 4. dormitories 5. fitness 6. refreshment area
7. changing room 8. circulations 9. parking 10. administration/data management
11. multipurpose computer room 12. lockers 13. courtyard

地下二层 second floor below ground

斜坡结构详图 slope structure detail

1. arrencada de mur prebeure esperes
2. reforç extrem solero segons detail
3. solera 20+15
4. terreny natural
5. cota fonamentació -0.50
cota superior solera -0.07

112
emergències

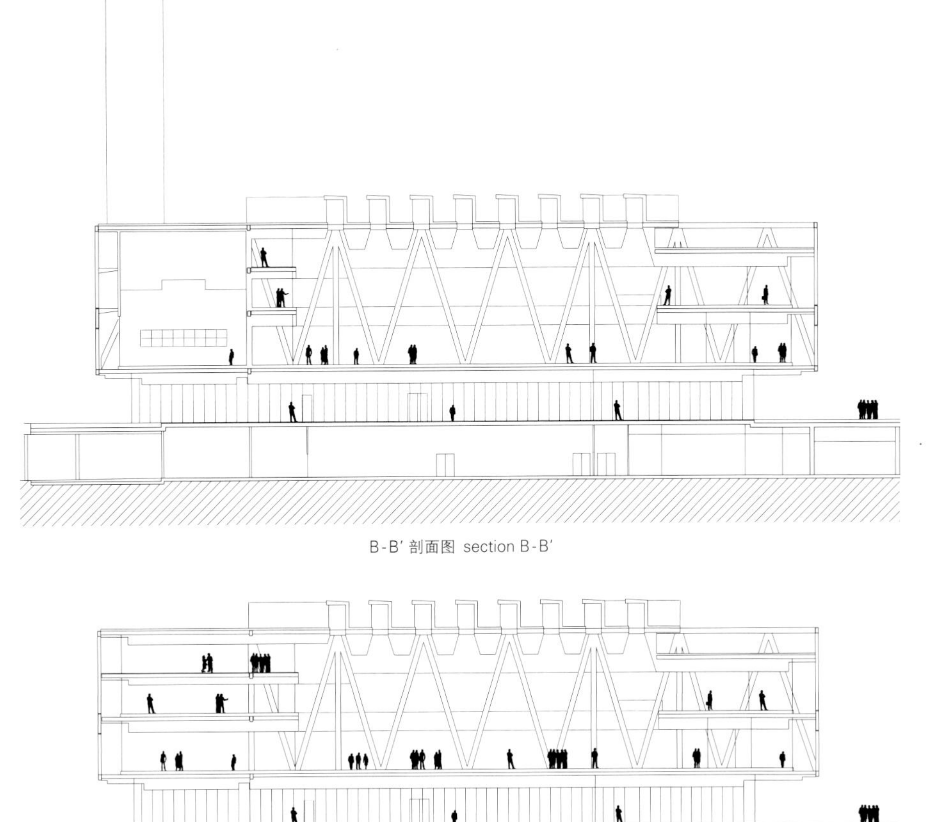

B-B′ 剖面图 section B-B′

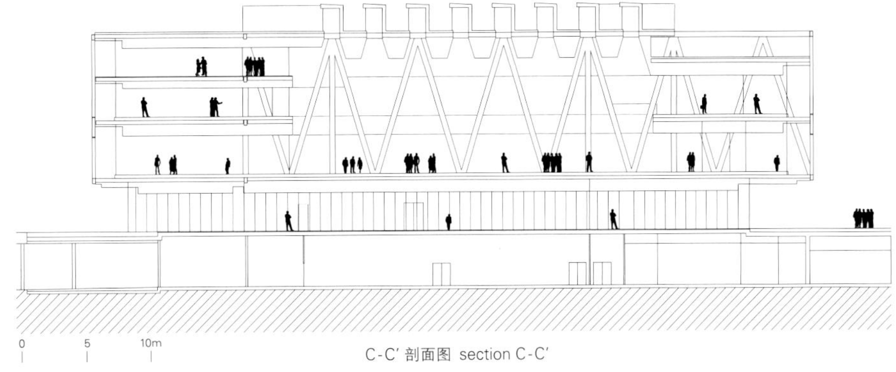

C-C′ 剖面图 section C-C′

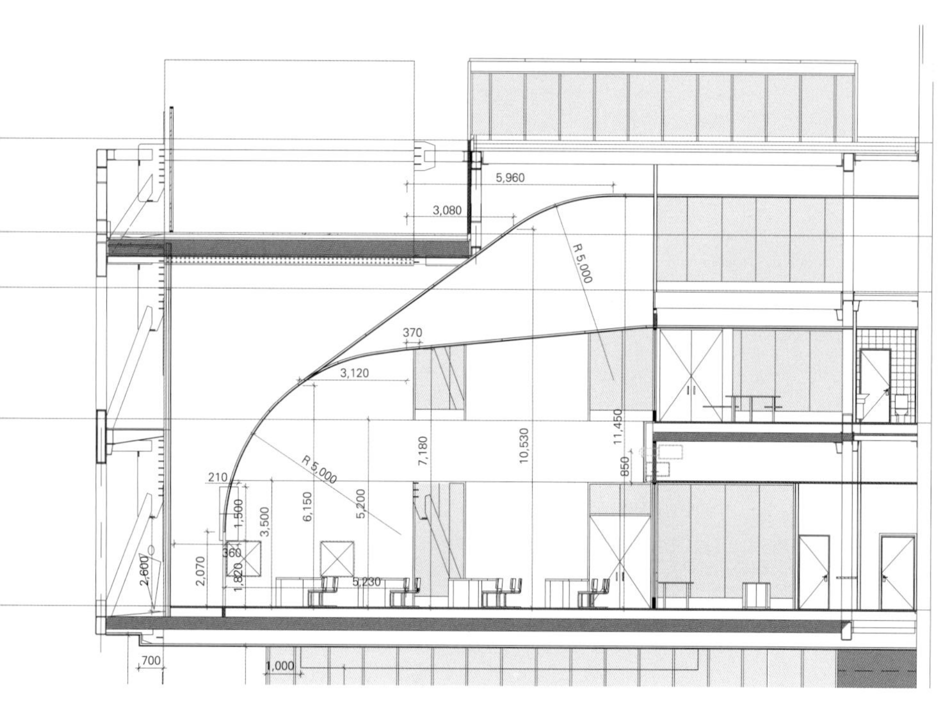

D-D′ 剖面图 section D-D′

E-E′ 剖面图 section E-E′

from a distance; now they communicate).

The building is horizontally divided into three layers: services plinth, public area and operational levels. The services plinth (parking lot, changing rooms, stockrooms, resting areas, building services) becomes the element that adapts to the peculiarities of the plot (topography and shape). Its roof offers back to the environment a landscaped area, which meets with the public part of the building (auditorium, press room, restaurant) on the first floor. This floor, situated above the olive-tree fields, grants excellent views over the territory, leading to the next floors: the operative box.

The operative box is formed by a large metal structure and a plastic mesh that achieves a double objective: avoiding direct solar radiation on the facade, on all sides, and cancelling out the idea of openings in the facade, thus increasing the territorial aspect of the building.

The metal structure, apart from solving the functional needs of the operational rooms, also allows flexibility to be incorporated for future floor distributions and promotes an image of unity of the operational bodies integrated in the new emergencies management model, through its solidity and its white colour, neutral to the uniforms of the operational bodies.

To boost the coordination and the synergies amongst the operators, the operation rooms are arranged around a large space that also allows natural diffused sunlight to the inside of the operational box.

Otherwise, vertical communications are defined by four nuclei: the telecommunications tower nucleus – which is also useful when receiving visits – the authorities' one, the maintenance one and the main access for workers one.

This functional matrix (horizontal layers and vertical communications) coincides with the structural and conceptual configuration of the building: a ground floor (services plinth) with a reticular flooring and a metal structure box, that forms the operative area and is supported by the four concrete nuclei.

In order to protect operations, the building has high security measures, both exterior and interior. The need to work round the clock, 24 hours a day, 365 days a year, calls for the main building services (electricity, telecommunications, air conditioning) to be redundant and for the building to be able to be self-sufficient for at least 5 days, should the supply fail. ACXT Architects

萨格勒布紧急服务中心

Katušić Kocbek Arhitekti/Produkcija 004

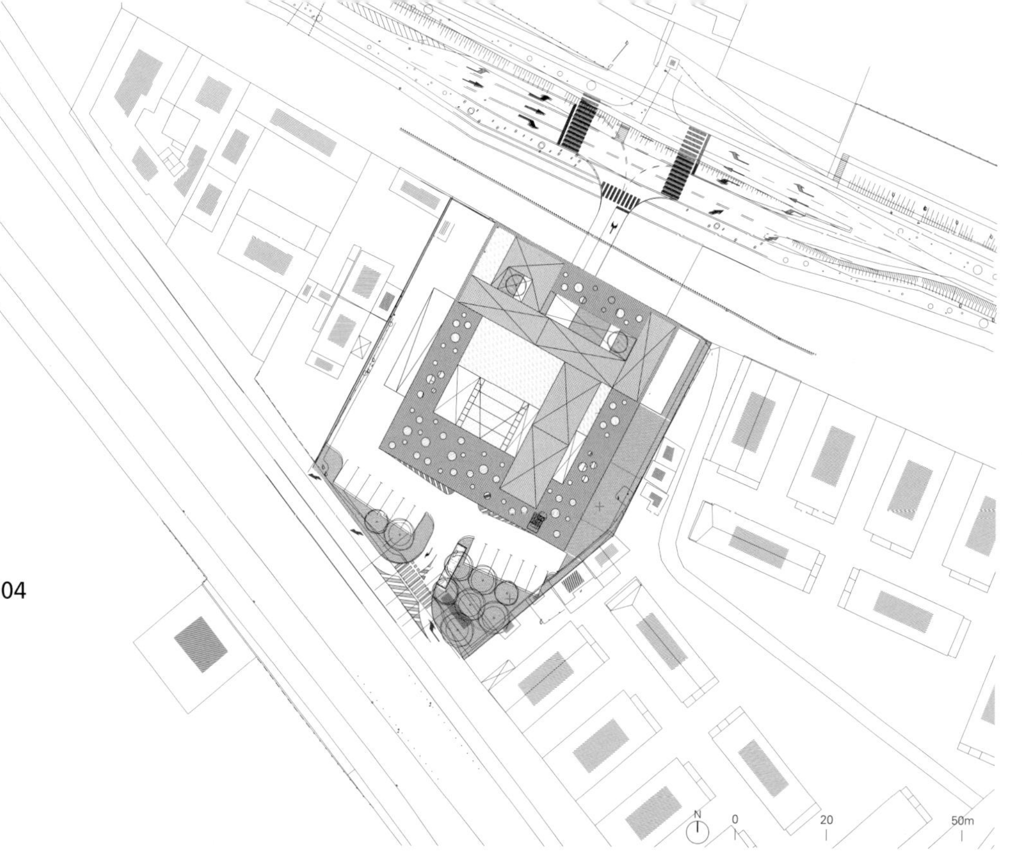

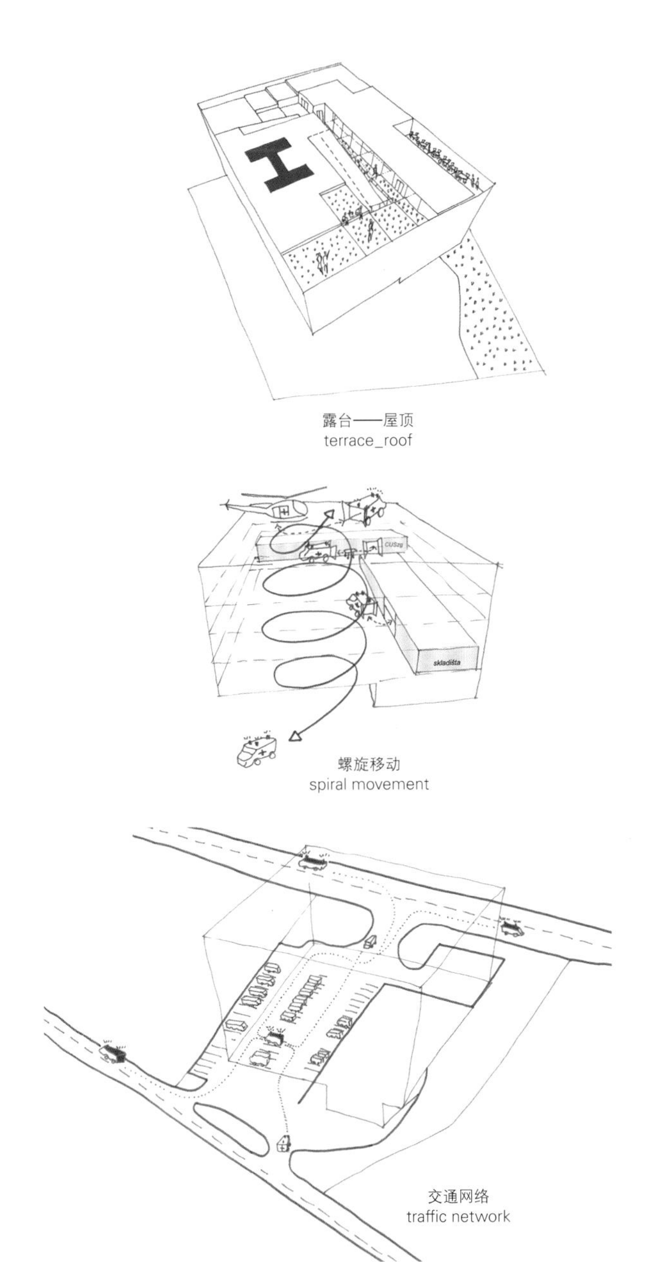

露台——屋顶
terrace_roof

螺旋移动
spiral movement

交通网络
traffic network

这座紧急服务中心将几种建筑类型（医院、诊所、车库、行政与教育大楼）综合起来，作为一种安全、能力和速度的新式城市标志应运而生。这座八层的设施包含了接待处、远程服务中心、康复诊所、门诊、萨格勒布市的药品供应库、演讲大厅、几间教室、实验室、行政服务部门、技术服务部门以及可容纳170辆医疗车辆的多层车库。该结构符合现代建筑的高标准，其表面积大约为14 000m²，可以为两百万市民提供服务。设有承重天花板的钢架结构符合建筑在亮度和空间多样性方面的需求。针对建筑大面积的立面而采用的有效合理的解决方案是薄膜网结构。

在多样化的周围环境中，材料（主要为纺织薄膜）的半透明特质在白天和夜晚会使人们对该建筑产生不同的感知。在白天，大楼的体量感在周围环境中格外突出，其简洁的白色极具吸引力。夜晚华灯初上的时候，建筑的膜立面会将整座大楼转变成一个大型的白色灯笼——这个白色的立方体就成了一个发光的地标性建筑。

Zagreb Emergency Terminal

Through a synthesis of architectural genres (hospital, clinic, garage, administrative, and educational buildings), the Emergency Terminal emerges as a new urban sign of safety, competence and speed. The 8-storey facility includes a reception and telecommunications service, resuscitation clinic, inpatient clinic, Zagreb City's medical supply storage, a lecture hall and classrooms, a laboratory, administration service, technical service and multi-storey garage accommodating 170 medical vehicles. Measuring up to the high criteria of contemporary architecture, the institution, approximately 14,000m² in surface area, can serve 2 million citizens. The steel skeletal construction with supported ceiling plates meets the architectural demands of lightness and spatial variability. An effective and rational solution for the large surface of the facade is the membrane net structure.

Within the heterogeneous surrounding the translucent character of the material (textile membrane - precontraint) contributes to the varying perception of the building during the day versus during the night. In the daylight the house dominates with its volume and attracts with the cleanness of white. When lit up at night the membrane facade turns the entire house into a large, white lantern – the white cube becomes an illuminated landmark.

Katušić Kocbek Arhitekti / Produkcija 004

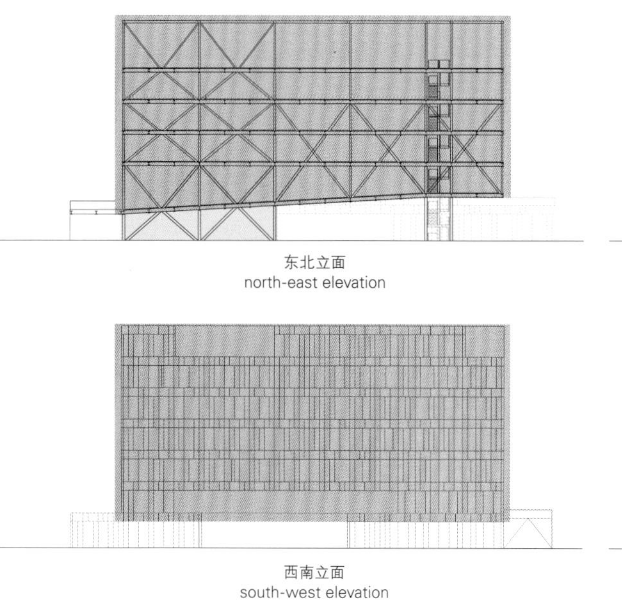

东北立面
north-east elevation

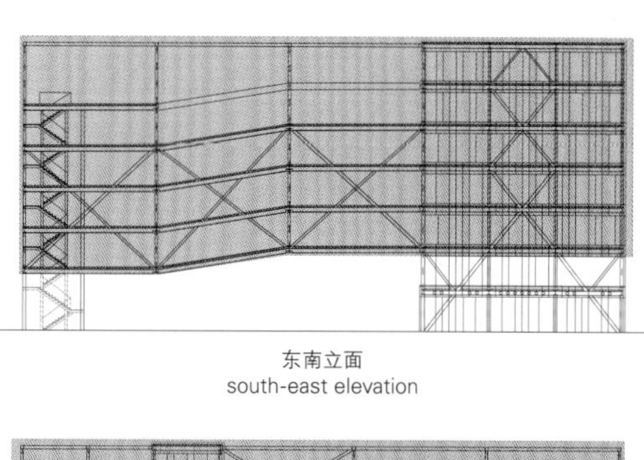

东南立面
south-east elevation

西南立面
south-west elevation

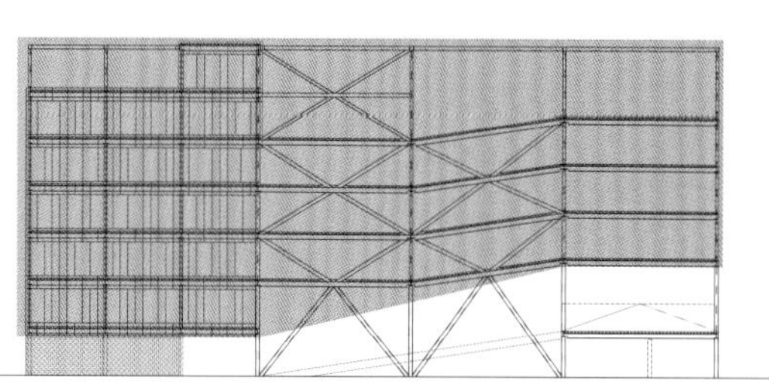

西北立面
north-west elevation

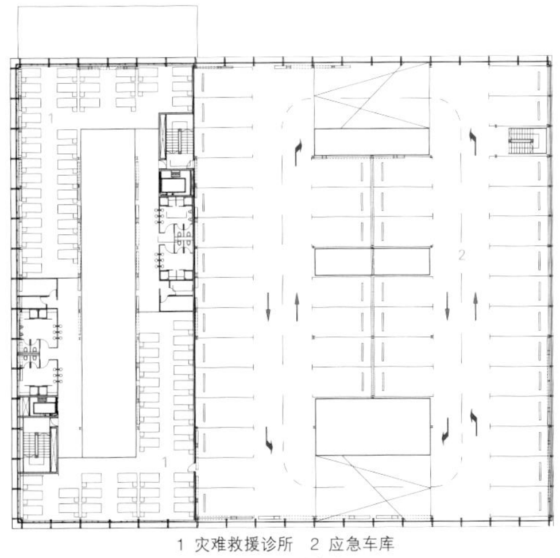

1 灾难救援诊所 2 应急车库
1. patient clinic for disaster relief 2. emergency garage

五层 fifth floor

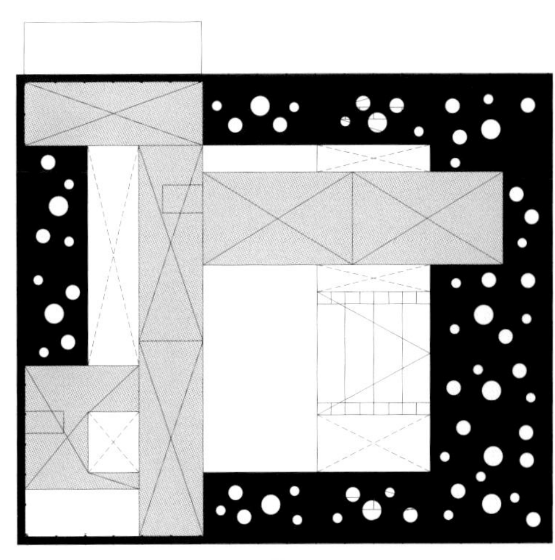

屋顶 roof

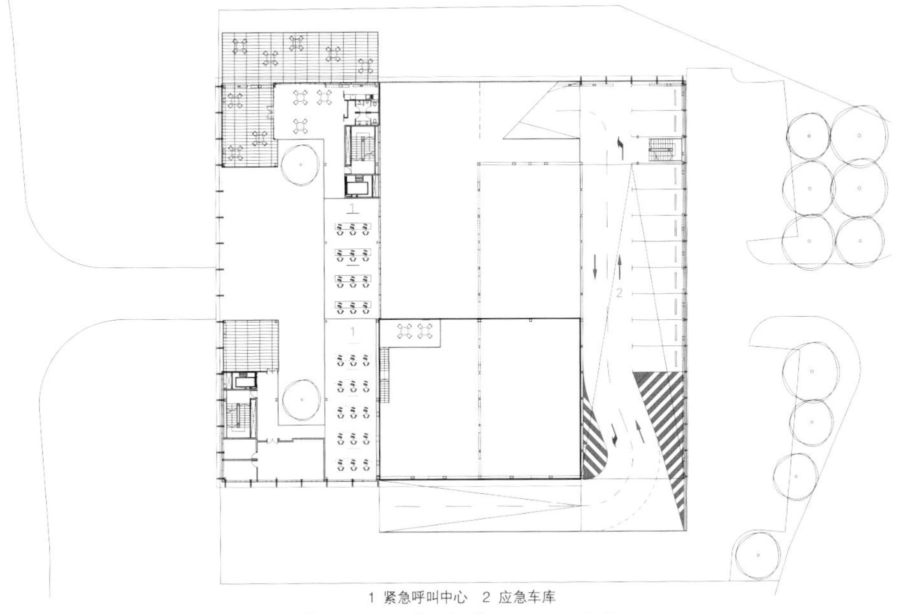

1 紧急呼叫中心 2 应急车库
1. emergency call center 2. emergency garage

二层 second floor

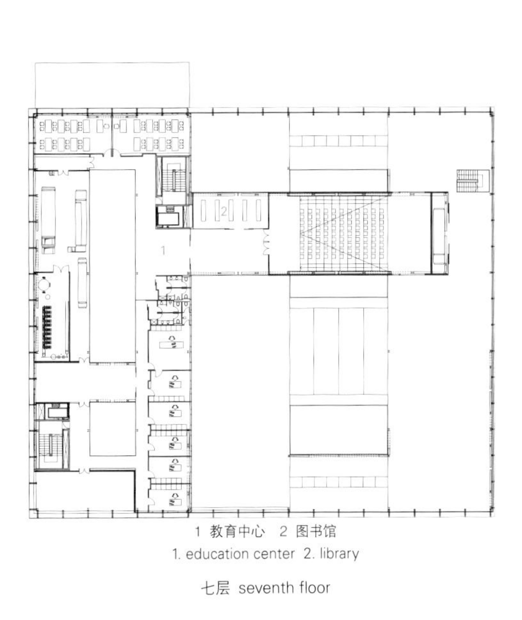

1 教育中心 2 图书馆
1. education center 2. library

七层 seventh floor

1 救护车 2 主要入口 3 有毒废物 4 应急停车场 5 汽车维修车间 6 洗车区
1. ambulance 2. main entrance 3. toxic waste 4. emergency parking 5. car repair workshop 6. car wash

一层 first floor

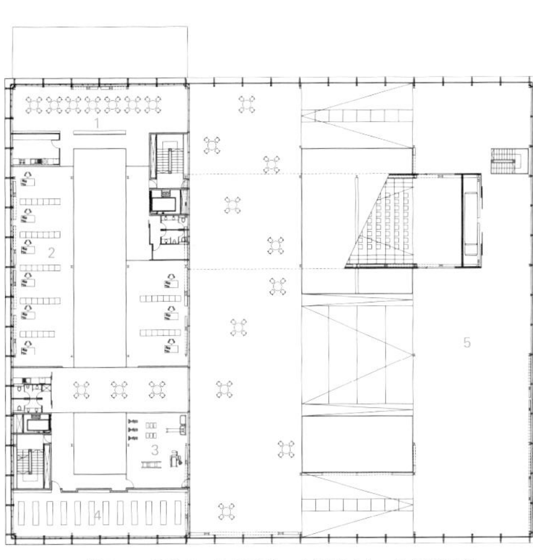

1 餐厅 2 行政区 3 档案室 4 健身中心 5 屋顶露台
1. restaurant 2. administration 3. archive
4. fitness center 5. roof terrace

六层 sixth floor

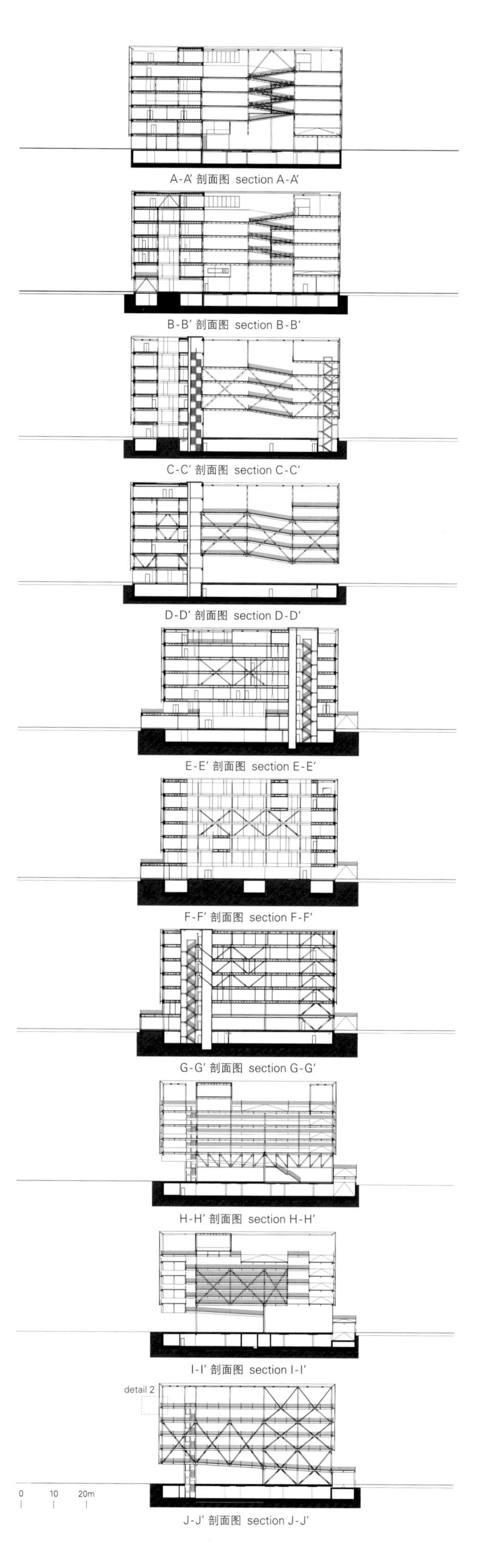

A-A' 剖面图 section A-A'

B-B' 剖面图 section B-B'

C-C' 剖面图 section C-C'

D-D' 剖面图 section D-D'

E-E' 剖面图 section E-E'

F-F' 剖面图 section F-F'

G-G' 剖面图 section G-G'

H-H' 剖面图 section H-H'

I-I' 剖面图 section I-I'

J-J' 剖面图 section J-J'

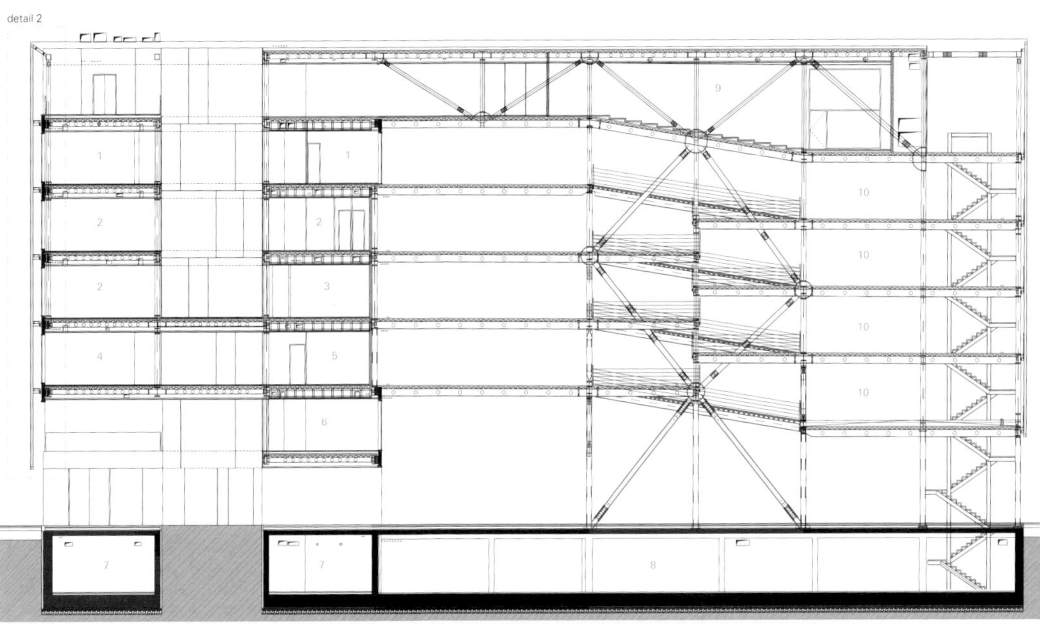

1. administration 2. patient clinic for disaster relief 3. medical equipment 4. emergency teams 5. apparel hold
6. emergency call 7. employees wardrobe 8. employees garage 9. auditorium(education center) 10. emergency garage

K-K' 剖面图 section K-K'

项目名称：Zagreb Emergency Terminal
地点：Zagreb, Croatia
建筑师：Davor Katušić
项目建筑师：Martina Ljubičić
项目团队：Margareta Ćurić, Jana Kocbek, Robert Franjo, Ivo Petrić, Marija Burmas
造型与标志设计：Juri Armanda, Karl Geisler, Bor Dizdar
建筑商：Dalekovod d.d., ZagrebMontaža d.d.
建筑面积：15,000m²
成本：EUR 15 million
竣工时间：2009.4
摄影师：©Studio HRG(courtesy of the architect)-p.116, p.117
©Miljenko Bernfest(courtesy of the architect)-p.118, p.121[top], p.122
©Željko Stojanović(courtesy of the architect)-p.120, p.121[bottom]

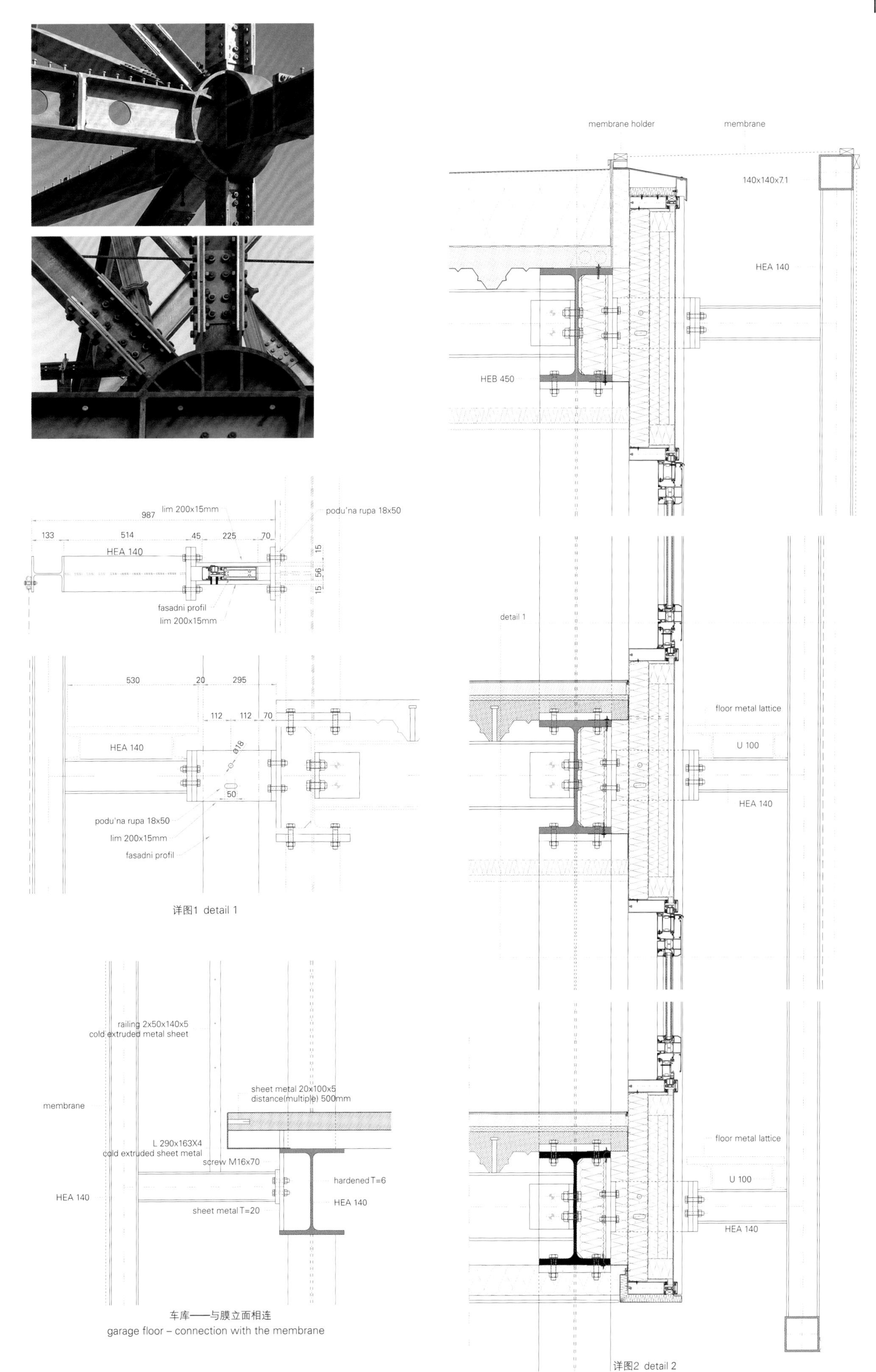

详图1 detail 1

车库——与膜立面相连
garage floor – connection with the membrane

详图2 detail 2

圣丹尼斯警察局

X-TU Architects

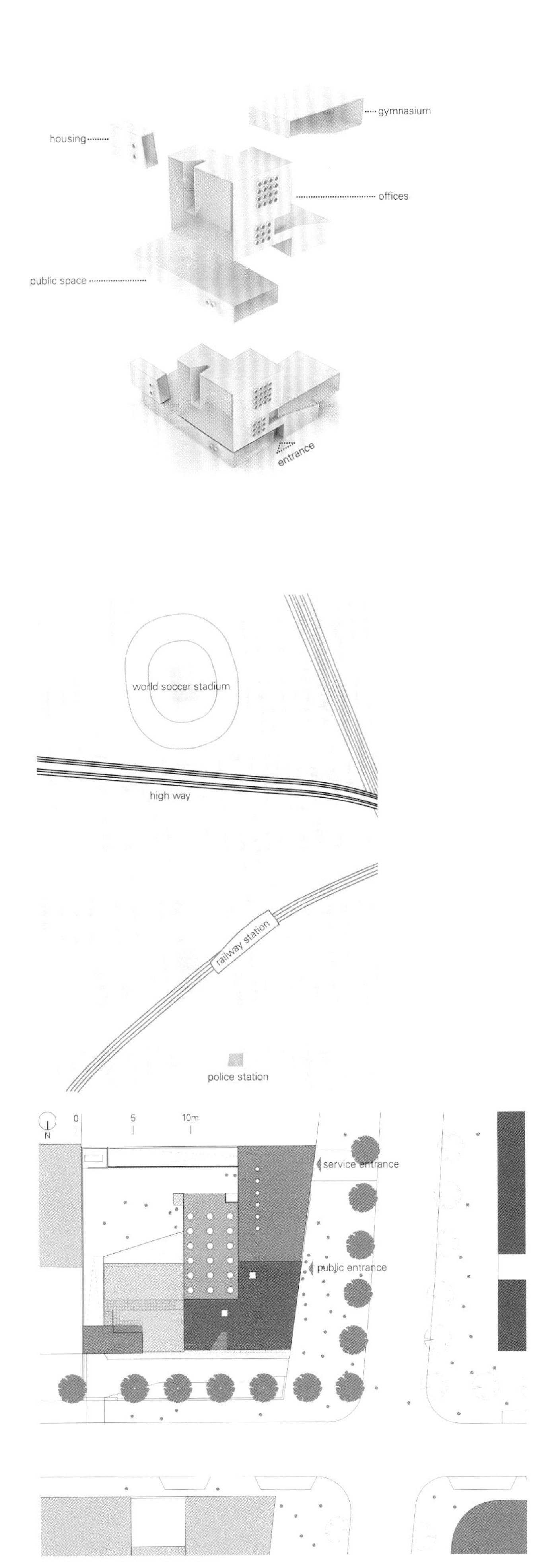

考虑到缩水后的项目规模和地块所处的位置，设计者的想法是将大部分项目构件都安插到一栋竖直、高耸的建筑中。设在街道转角处的这一建筑将成为一座标志性高楼。"特殊"的项目构件将会拥有自己的体量。它们借助玻璃幕墙将建筑结构从地块的一端拉伸到另一端。这些空间囊括了一层和二层，构成一个基座，其他部分均建立于其上。

底座沿着立面逐渐舒展开来，呈现出若干个内凹结构和凹口，嵌入或分割出不同的体量。这种互动式设计在各个体量之间建立起一种张力连接。这种雕塑式同时也是城市化的设计为未来周边建筑的落成做好了准备。它是一座由环境铸就的城市雕塑。

该开发项目深受周围圣丹尼斯平原（巴黎北郊）实际景观的启发。一片支离破碎的景象，集合了各类景观，与具有不同比例、各种材料构成的建筑群交叠在一起，构建起一片富有诗情画意的天地，有时美得令人惊叹，透过一道道空隙，天空几乎可以渗透到整个景观的内部。

立面的设计理念是在不透明与透明之间寻求永恒的碰撞。不透明墙体由光滑混凝土筑成。几排装饰有乳白色圆筒结构的舷窗贯穿了立面，在夜间可以照亮整个结构。根据室内设施所需机密程度的不同，玻璃墙上装配有不同类型的玻璃。

位于Landy大街塔楼内的办公室都装配有丝网印花玻璃，不透明与透明部分的特殊排列方式构成了一种波浪式的几何形状。这一结果促使设计者在室外安装了一块LCD显示屏，同时，当人们从室内向外看时可享受到一定程度的透明感，而从外面看时却有足够的遮蔽效果。宽大的玻璃表面让该建筑即便在夜里仍能保持自身的标志性地位，当人们聚拢到圣丹尼斯体育馆周围时，这座塔楼就变成了方圆几百里之内都可以看得到的发光标志。

底层分为两部分。第一部分是一片公共休息区，可以通往投诉办公室和楼上。第二部分是为值班保安专设的休息区。两个区域内的活动都在一台监视器的视线范围之内，监视器安装在一个柜台上，该柜台将两个休息区分隔开来。其中心位置便于观察警察和公众的出入情况，同时还可监控到"武器储藏室"、GAV区以及各个楼层的入口。

值班保安休息区一直延伸到二楼，在那里可以找到其他日常维护部门和健身房。这种延伸式设计放在休息区里貌似是一个错误，休息区上方是放置运动设备的大厅，休息区就从那里采光。运动设备所在的大厅同样有自己的入口——一段起自内庭的楼梯——这给了它更多的独立性，避免在法兰西体育场比赛期间干扰到警察局的其他部门，体育场产生的影响会大一些。

设计者对室内循环系统给予了特别的关注。这间警察局内包括了许多种服务部门，每一种都有自己的安全需求。因此，通向特别部门（负责政治安全的部门）的入口单独开设在Landy大街上。同样地，值班宿舍的入口设在后庭，与配楼的出口位于同一楼层。

Saint-Denis Police Station

Given the reduced program size and the location of the land parcel, the idea was to insert most of the program elements in a high vertical volume, which set on a street angle will stand as an emblematic tower. The "special" elements of the program will have volumes of their own. They strain the composition from one end of the parcel to the other with glass walls. These spaces contain the ground floor and the first level and create a pedestal from where the rest of the masses stand on.

The plinth unfolds itself along the elevation creating set backs and notches where the different volumes snap in or detach themselves from. This interaction creates a tension link between the volumes. The sculptural as well as urban solution, anticipates the future buildings in the surrounding parcels. It is an urban sculpture shaped by its environment.

This development is deeply inspired by the actual surrounding landscape of the Plaine Saint-Denis (northern Parisian outskirts).

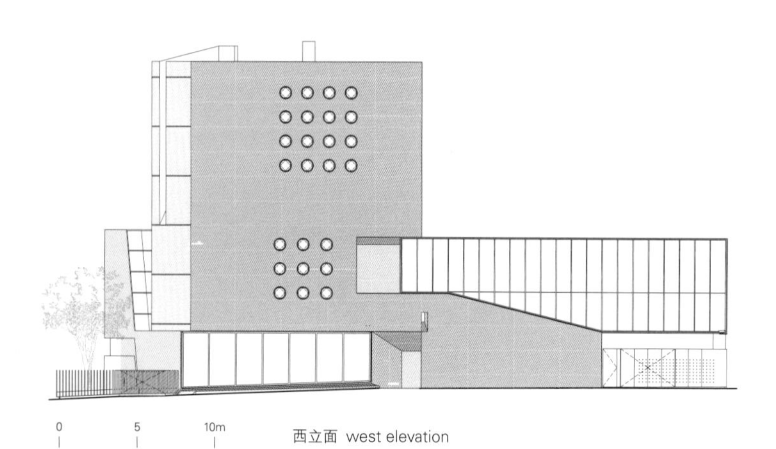

西立面 west elevation

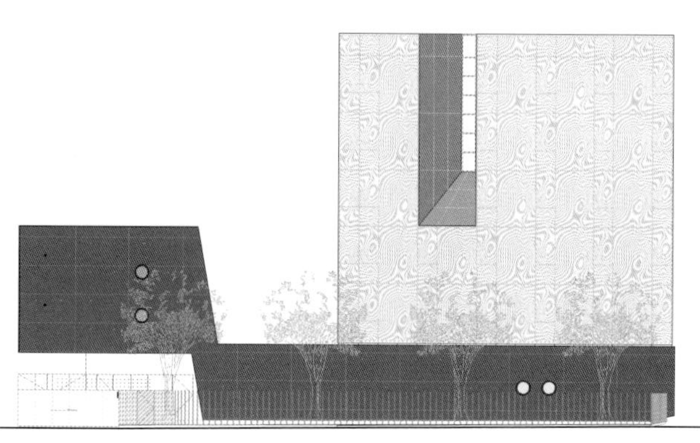

北立面 north elevation

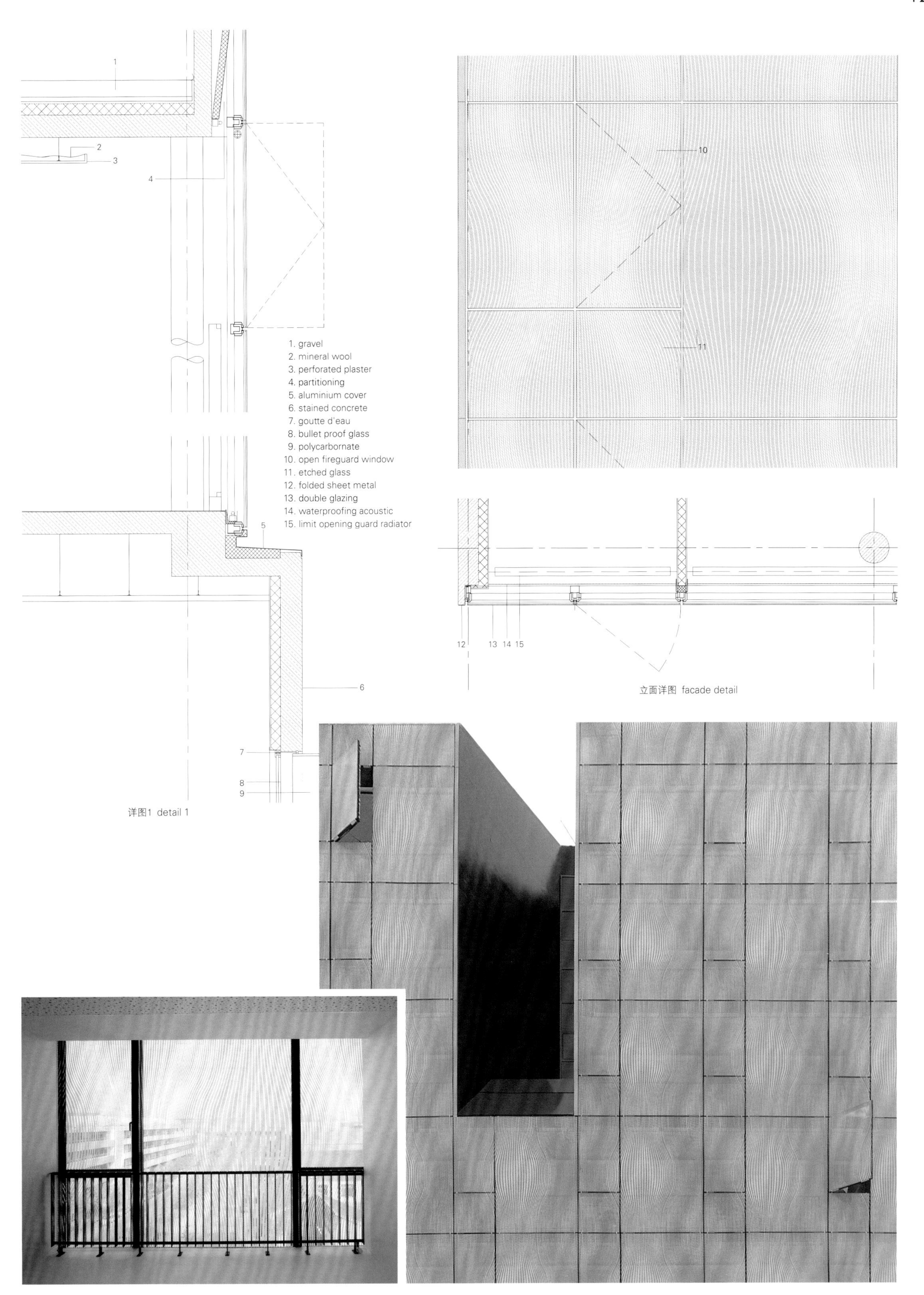

详图1 detail 1

立面详图 facade detail

项目名称：Saint-Denis Police Station
地点：Saint-Denis, France
建筑师：Anouk Legendre, Nicolas Desmazieres
项目团队：Roel Dehoorne, Gaelle Leborgne, Melanie Bury, Ingrid Manger
设备安装工程师：Francilia
安防工程师：Qualiconsult Securite
工程：ESTAIR
经济师：F. Labrousse
制图人员：Frederic Teschner
甲方：Ministery of interior
功能：office, security, fitness room
占地面积：1,622m² 建筑面积：5,109m² 总建筑面积：3,857m²
材料：serigraphed glass, black concrete, lacquered metal
造价：EUR 5,790,000
竣工时间：2009
摄影师：©JM Monthiers (courtesy of the architect)

A broken assembly overlapped with buildings of different proportions and materials, creating a poetic universe, sometimes astonishing, slashed by void making the sky invade almost all the landscape.

The elevations are conceived as constant opposition between opacity and transparency. The opaque walls are done with glossy concrete. Series of portholes dressed with opalescent cylinders punctuate the elevation and illuminate it at night. The glass walls are equipped with different types of glass depending on the confidentiality degree required by the facilities.

The offices in the tower on the Rue de Landy are equipped with silk screen printed glass, where the positioning of the transpar-

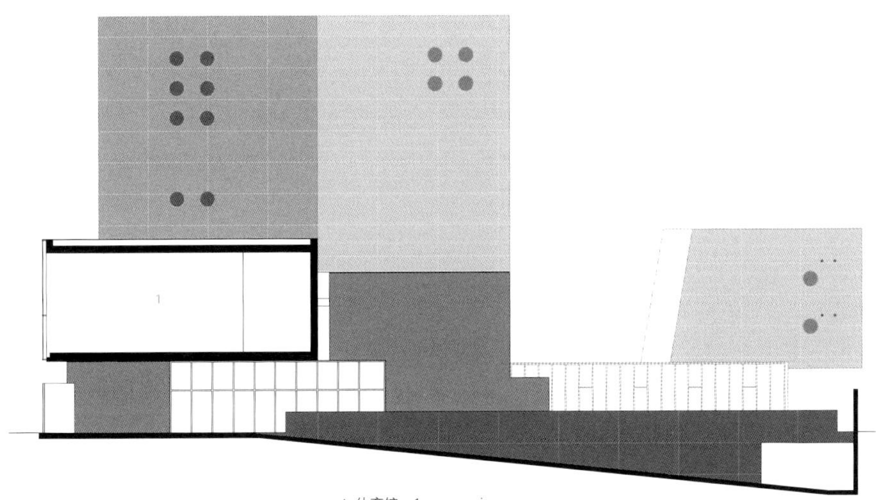

1 体育馆 1. gymnasium

A-A' 剖面图 section A-A'

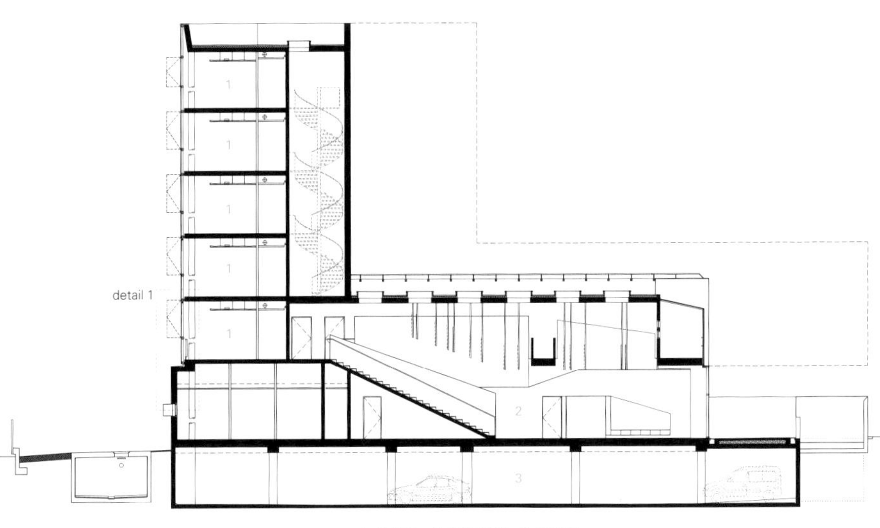

1 办公室 2 服务大堂 3 停车场

1. office 2. service lobby 3. parking

B-B' 剖面图 section B-B'

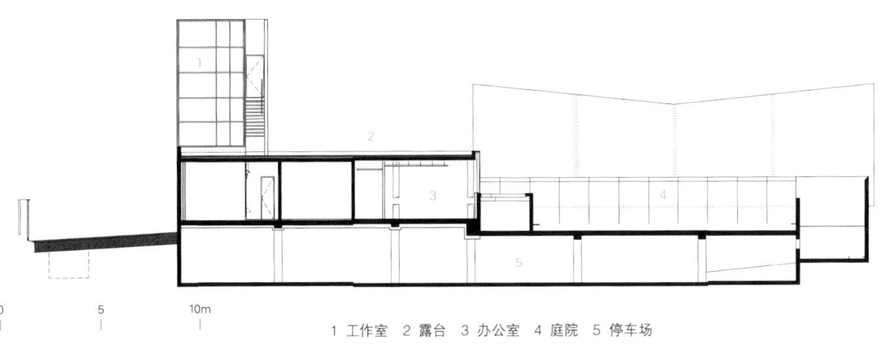

1 工作室 2 露台 3 办公室 4 庭院 5 停车场

1. studio 2. terrace 3. office 4. court 5. parking

C-C' 剖面图 section C-C'

ence and the opalescence create a wave type of geometry. This result which externally evokes a LCD screen also lets a certain transparency from the interior towards the outside, although giving enough opacity from the outside. The large surface of glass lets the building keep its emblematic role even at night, when the crowd gathers around the Saint Denis Stadium, the tower becomes a luminous signal seen from far away.

The ground level divides itself into two parts. The first is a public lobby giving access to complaint offices and accompanied access to superior levels. The second one becomes the lobby for the on-duty guards. The movement inside both lobbies can be observed by a supervisor who is installed in a counter which separates the two spaces. The central position of this counter makes it easy to visualize the comings and goings of the police and of the public, and also commands the access to the "weapon storage" , to the (GAV) zone, and to the floors.

The on-duty lobby extends to the first level where can be found the rest of the daily maintenance rooms as well as the fitness room. This extension appears as a fault in the lobby's space which brings light from above provided by the hall servicing the sport facilities. The sport facility also has its own access, a staircase starting from the inner court, giving it more independence, avoiding the rest of the station to be disturbed during match days at the Stade de France, where the influence can be greater.

Special attention was brought to the internal circulation. This police station contains many types of services each having their own degree of security requirement. Therefore the access to the Special Branch (branch for political security) is done by an additional entrance located on Rue de Landy. In a same way the on-duty housings are accessible from the back court situated on the same level as the annex exit. X-TU Architects

四层 fourth floor

三层 third floor

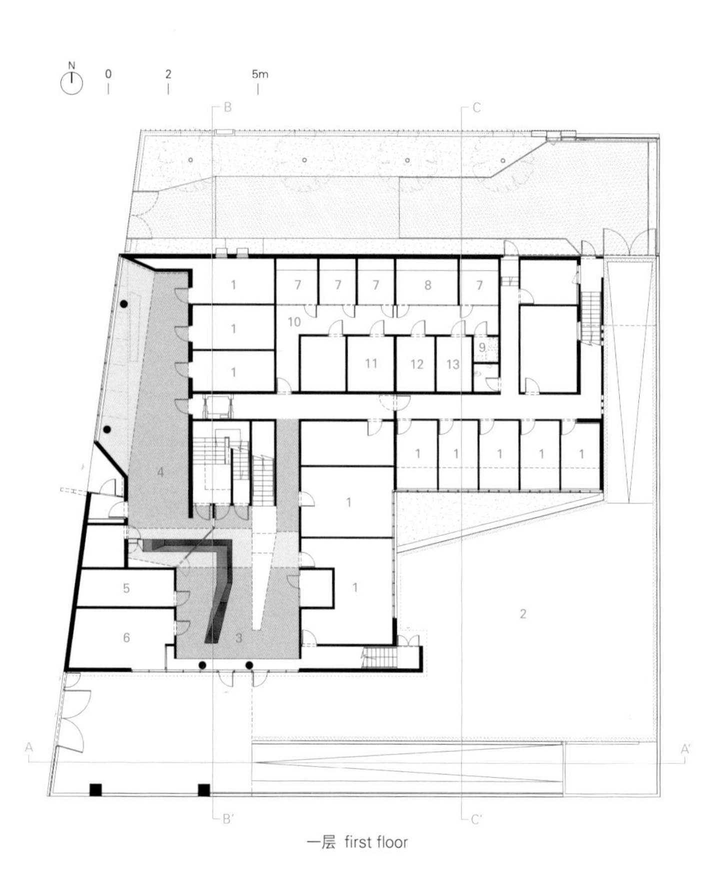

一层 first floor

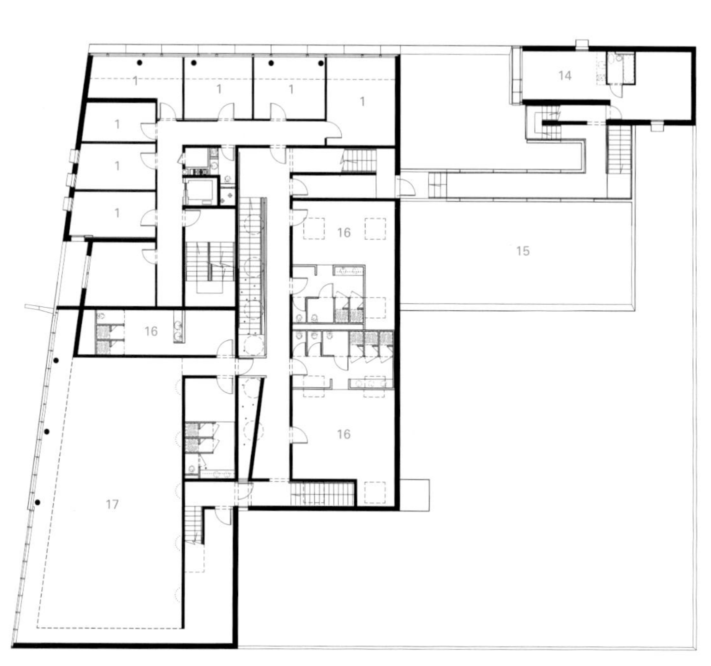

二层 second floor

塞维利亚的警察局

Paredes Pedrosa Arquitectos

塞维利亚的新警察局是一座独立式建筑，它靠在场地的东南角，朝向Emilio Lemos环形街开放。建筑的外形如同一个不规则的四角星，是受全景模型的启发而设计出来的。该建筑被风景优美的条状地块围绕着。这条天然的带子一直延伸到外部透明的篱笆处，并自然地贴近建筑的翼楼。这样，设计出的体量就获得了相当长的立面和尽可能多的房间，而这些房间统一朝翼楼间的外部空间开放。

警察局周围有很多高层住宅楼。相比之下，警察局反倒选择了一种水平延展型的布局形式，但又不紧贴地面，从而形成了该建筑在城市中独具代表性的形象。

外立面分为不同的两层。上层是连续不断的悬挑式白色混凝土墙，将建筑四周围合起来。这层表皮具有褶皱状的外观，看起来像是从顶板上悬下来的连续不断的窗帘。这层固体遮光罩的下面是玻璃覆层，它可以根据太阳方位不断变化，提供渐变的保护，为室内遮挡强烈的阳光。铝质框架内镶有两层安全玻璃，内层玻璃为建筑外部展示出一种银色的饰面。

该项目被设置在星星的四个角内。公共区域紧挨着入口，位于一层透明的表皮后面，为到访警察局的市民传达一种亲近感。这种建筑类型与它的行政用途高度相关。该项目分为两层进行开发，其中一层部分建在地下。

绿色的斜坡造就了警察局的外观，也保护了容纳室内停车场和牢房的地下层。地下层又分为两个主要区域：一个被设计为讲授见义勇为、法律救助和未成年人保护等主题公共区域，另一个是关押被逮捕的人的封闭区。档案室、存储室和更衣室也位于这一层，自然光可以通过一座狭长的庭院洒落进来。

在一层，四角星的形状使得中央可以有一个将项目分配到不同翼楼中的入口。这一入口由天井一采光井（将公共和室内区域分隔开）竖向切断。检查通行证和身份证的区域位于大厅右侧，而正式申诉区则位于左侧翼楼。另外两个翼楼包含不同部门的行政办公室。

在中央空间内，公共楼梯的存在不容忽视，它将周围四座翼楼集中起来，同时由于自然光的入射，它也将上下两层联系到一起。另一个较为私密的楼梯系统还包括一部电梯，位于停车场和牢房中间。

由于采用了白色石灰岩路面和格状金属板构成的模块天花板，一层具有良好的空间连续性。

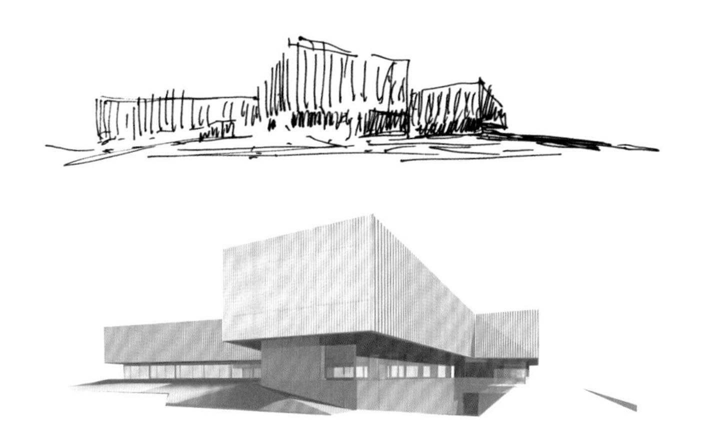

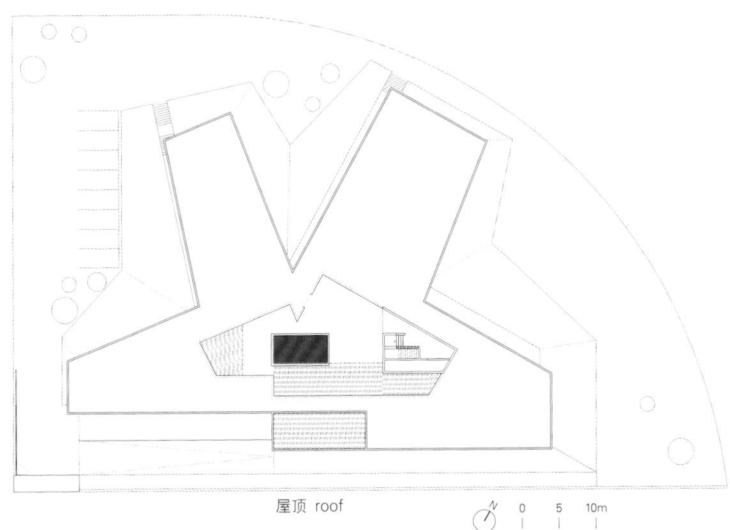

屋顶 roof

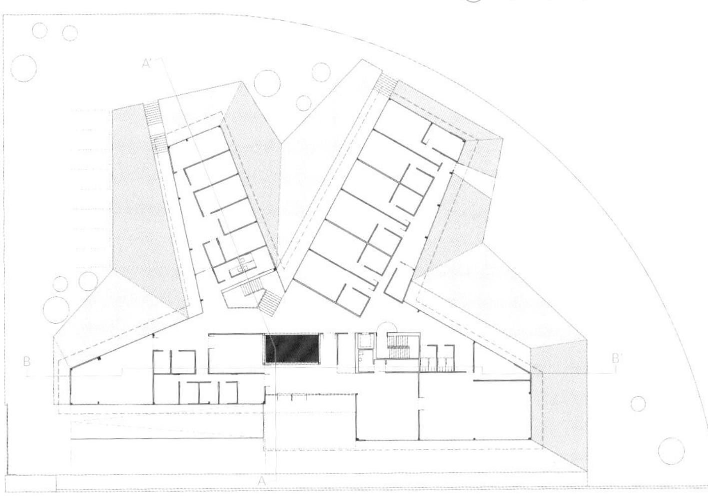

一层 first floor

Police Station in Seville

The new police station in Seville is a detached construction that leans on the southeast edge of the site and opens up towards the circular street Emilio Lemos. The building is shaped as an irregular four-pointed star, inspired in a panoptical model. A landscaped strip then surrounds the construction. This natural belt is extended up to the exterior transparent fence and naturally embraces the wings of the building. In this way, the proposed volume achieves an extensive facade with a maximum number of rooms opening towards the outdoor spaces in between the wings.

A considerable number of high-rise dwelling buildings surround the police station. In contrast, the project opts for a horizontal and extended organization that does not seem to touch the ground and shapes the building's representative image in the city.

The external facade is composed in two differentiated levels. Above, a continuous suspended wall made of white concrete wraps the perimeter of the building. This skin has a corrugated appearance and looks like an uninterrupted curtain hanging from the superior slab. Beneath this solid eyeshade, a glass wrapping shelters the interior from the harsh sunlight thanks to a gradual protection that adapts itself according to changing orientations. The aluminium frames hold a security glass, made up of two layers, the second of which exhibits a silver finishing of the exterior.

The program is contained within the four wings of the star. The public areas are located right next to the entrance, behind a transparent skin that conveys an appearance of proximity to the citizen who approaches the police station. The building typology is highly adapted to its administrative use. The program is developed on two levels, one of which is partially built underground.

Green slopes that shape the external appearance of the police station protect the underground level, which hosts the internal parking area and cells. This level is in turn divided into two main areas: one is designated for public processes like identity parades, legal assistance, or under 18s security and the other is an enclosed area for those under arrest. Archives, deposits and changing rooms are also located on this floor, provided with natural light through a lineal courtyard.

On the ground floor, the four-pointed star shape permits a central access that distributes the program among the different wings. This entrance is crossed vertically by a patio-lightwell that separates public and internal areas. The area for passports and IDs is located on the right of the lobby and the one for formal complaints is on the left wing. The other two wings host the administration offices for different departments.

In the central space, the public stairs gain a very strong presence, centralizing the four surrounding wings and uniting the two levels thanks to the invasion of natural light. The other staircase, more private, includes a lift and is located between parking area and cells.

The spatial continuity on the ground floor is achieved thanks to the white limestone pavement and the modular ceiling made in grated metallic plates. Paredes Pedrosa Arquitectos

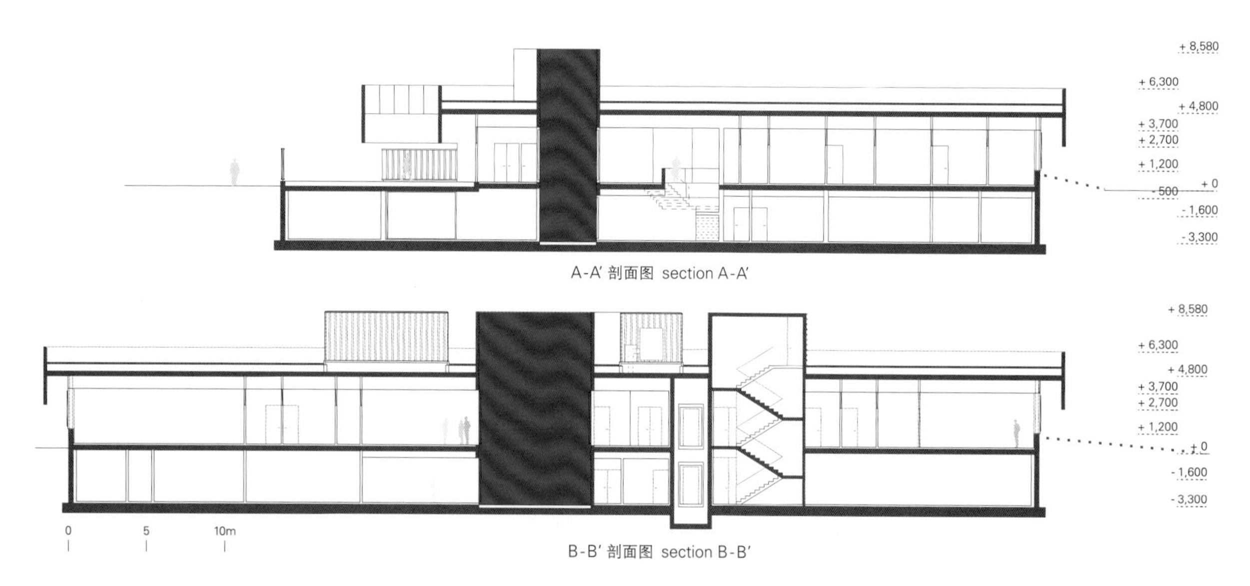

A-A' 剖面图 section A-A'

B-B' 剖面图 section B-B'

项目名称：Police Station in Seville
地点：Avenida Emilio Lemos / Calle Médicos Mundi. Seville
建筑师：Angela García de Paredes. Ignacio Pedrosa
项目团队：Álvaro Oliver, Álvaro Rábano, Clemens Eichner, Eva Urquijo, Djamila Hempel, Jorge López
场地建筑师：Ángela García de Paredes, Ignacio Pedrosa, Álvaro Oliver
结构工程师：GOGAITE S.L.
机械工程师：GEASYT S.A
技术管理：Luis Calvo, Gonzalo Cátedra
承包商：JOCA Ingeniería y Construcción S.A.
甲方：Gerencia de Infraestructuras de la Seguridad del Estado
用途：police station, offices
楼面面积：3,177.93m^2
设计时间：2006
施工时间：2008—2009
造价：EUR 2,858,310.66
摄影师：©Luis Asin(courtesy of architect)-p.134, p.135
©Roland Halbe-p.132

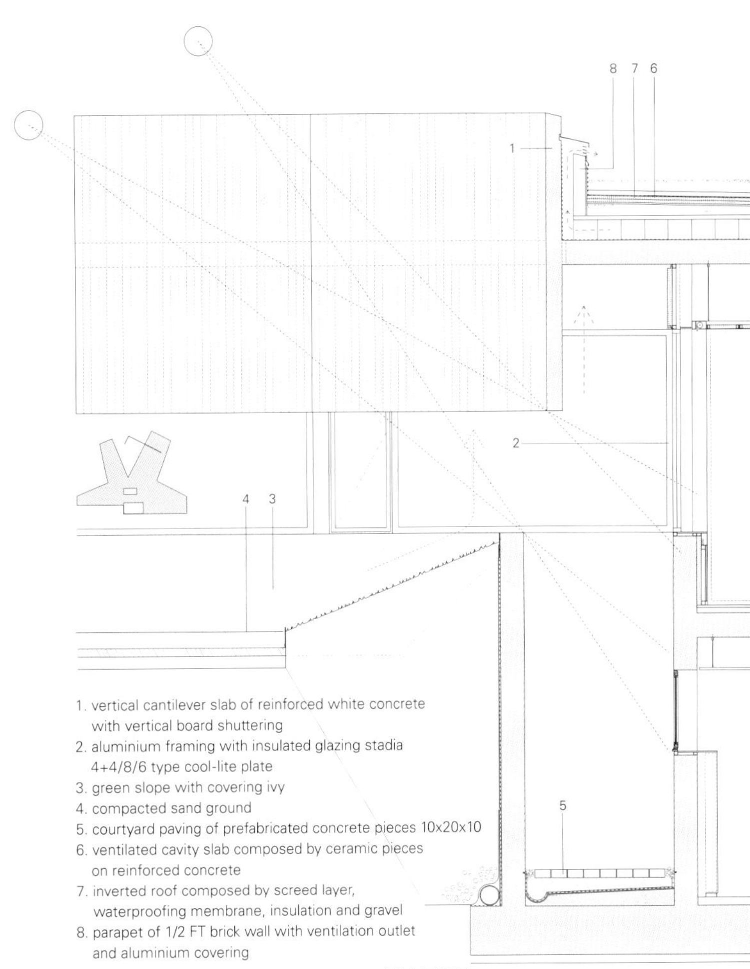

庭院剖面详图
courtyard section detail

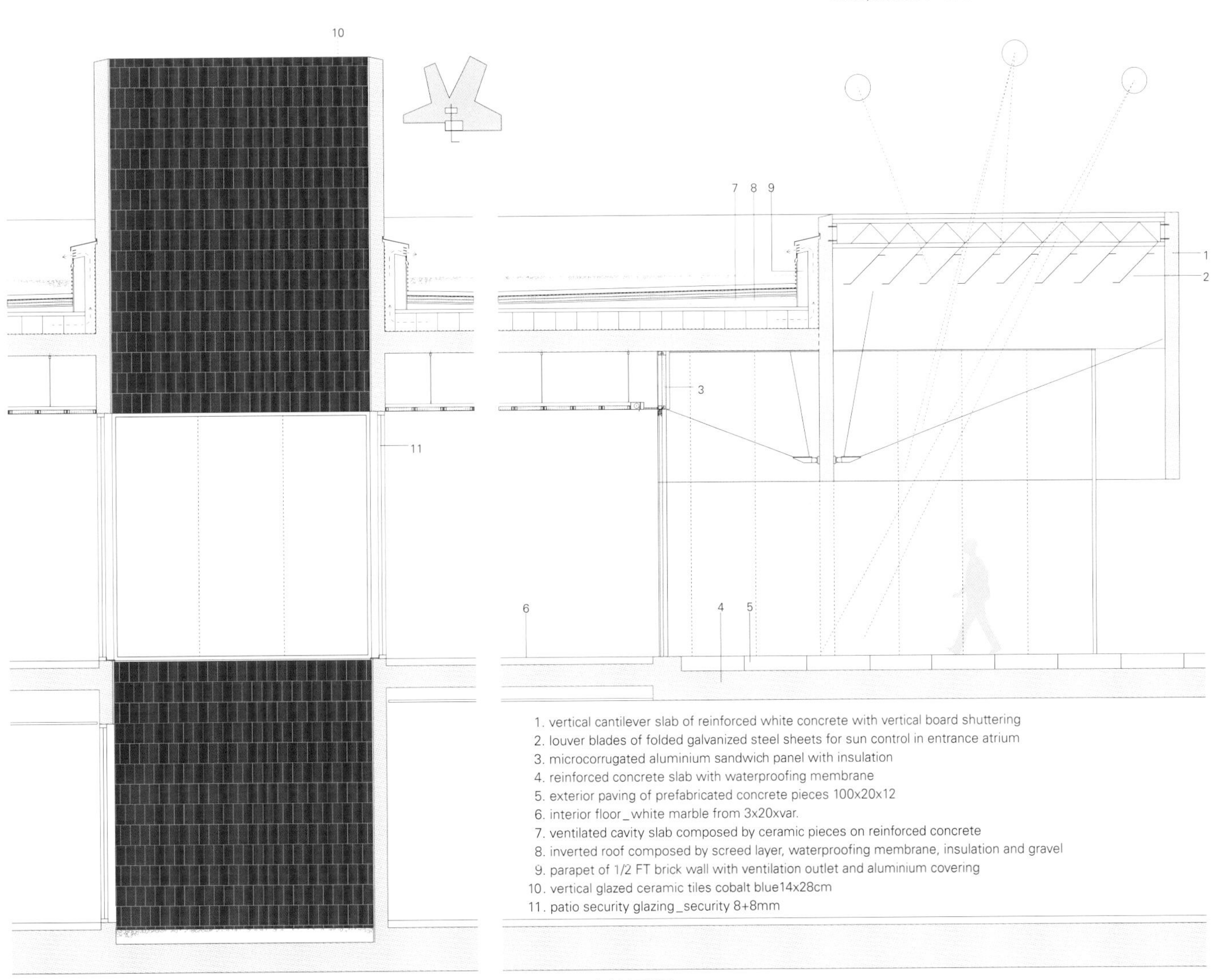

中庭剖面详图
atrium section detail

曼萨纳雷斯的警察局

Estudio Lamazeta

曼萨纳雷斯的新警察局建在一家旧酿酒厂的场地上，是一次建筑设计竞赛的结果，而该竞赛的最初目的是要建一座博物馆。就在建筑完成了80%的时候，政府决定将该建筑改成一所警察局。

该建筑的周边环境混杂，有体育建筑、石灰墙和开放空间；一系列抽象而有力的体量将建筑围合起来，与外界隔绝，而各个体量由狭长的庭院分隔开；通过庭院，人们可以看到建筑的中央空间。

人们可以通过一个带顶的庭院进入建筑，该庭院为既是等候区又为接待处这一中央空间带来光线和热量。从中央空间出发，人们可以抵达不同的办公室，也可以抵达那些狭长的庭院。两个小型的蓝色盒状房间最初被设计为展览室，在房间内独立排布。它们试图展现建筑在建造过程中经历的用途转变。

为了获得适当的自然光照，该建筑设有大型天窗，这是由建筑最初的用途决定的。然而，为了保证办公区的通风效果，采光和通风窗开在不同的狭长庭院中，而不是在立面上，所以外部仍然保留了最初浑圆的形象。

在建筑的外部饰面上，建筑师采用了白色的陶瓷贴面，将大小和比例不一的部分连接起来。经过设计，这种拼接式图案在纹理和色彩略有差异的立面的不同部分逐渐消失。

Manzanares Police Station

Located on the site of an old winery, the new Police Station of Manzanares is the result of an architectural competition initially announced for the construction of a museum. When the construction was 80% completed, the government decided to adapt the building to the current use.

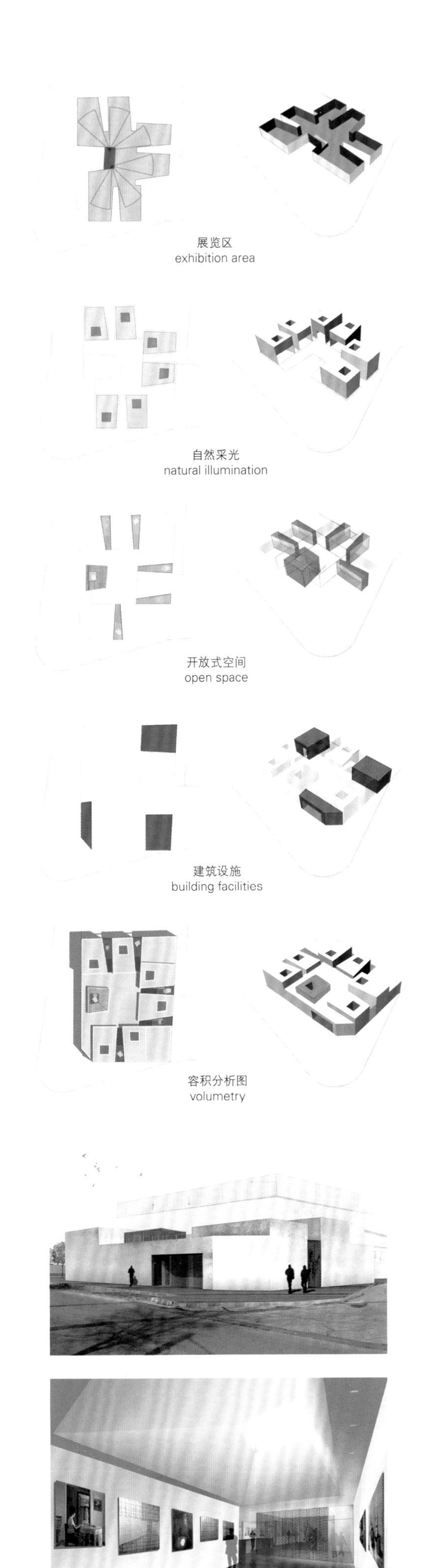

展览区
exhibition area

自然采光
natural illumination

开放式空间
open space

建筑设施
building facilities

容积分析图
volumetry

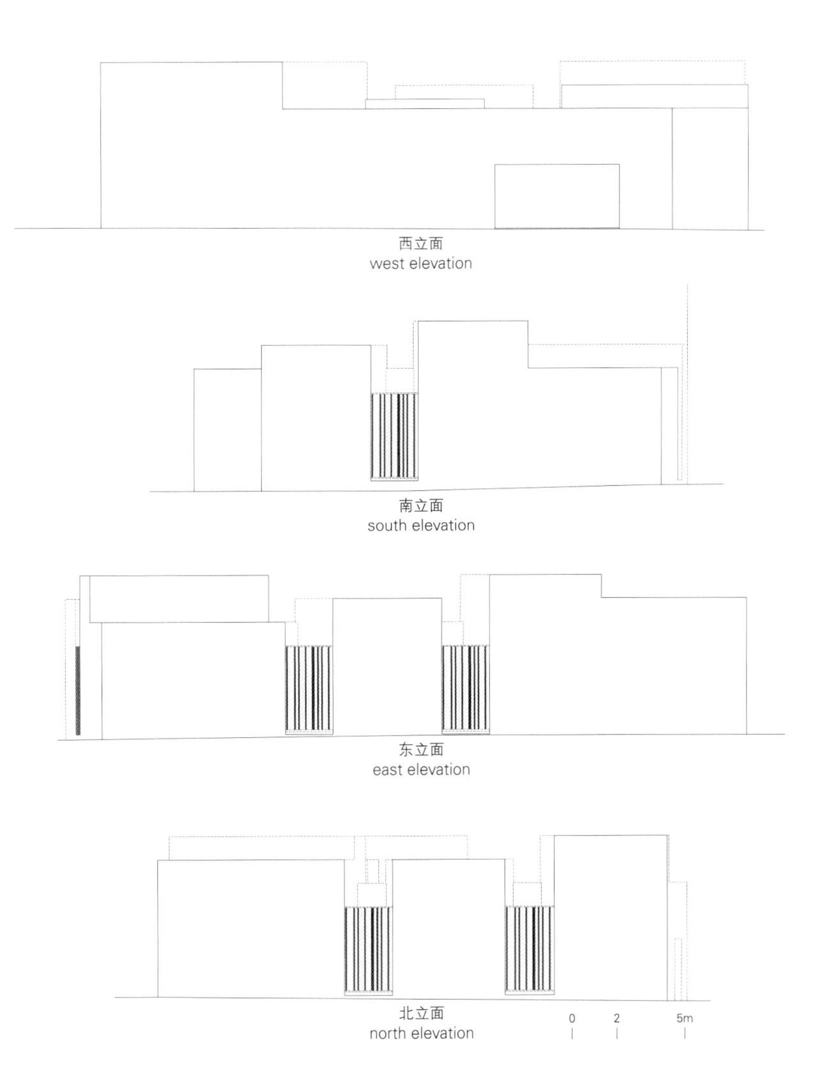

西立面
west elevation

南立面
south elevation

东立面
east elevation

北立面
north elevation

Surrounded by a nondescript environment like sport buildings, lime walls and open spaces, the building is enclosed to the outside with a set of abstract and strong volumes. The different volumes are separated from each other by elongated courts that let glimpse into the central space of the building.
The entrance is through a covered court that brings light and warmth to the central space, which is both waiting room and reception. From this central space you can have access to various offices, as well as to the different elongated courts. Two rooms seem like little blue boxes installed as isolated objects within the rooms that were originally designed as exhibition rooms. They are trying to show the change of use occurred during the construction of the building.
To achieve an appropriate natural lighting, large skylights have been built, due to the use that the building originally would have had. However, to ensure the ventilation of the offices, lighting and ventilation windows have been opened in different elongated courts, not in the facade, so the outside rotund image remains as the original.
For the outside finishes of the building we have used a white ceramic tiling that combines pieces of different sizes and proportions. The patchwork designed disappears gradually in different parts of the facade where the texture and color used are slightly different. Estudio Lamazeta

项目名称：Manzanares Police Station
地点：Antigua Carretera de Madrid, Manzanares (Ciudad Real)
建筑师：Lucía Manresa, David Mata, María Moreno
合作者：Enrique Salazar, Marta Ivars, Javier Jimenez, Belén Jurado
建筑团队：Construcciones Anclade SL
甲方：Excelentísimo Ayuntamiento de Manzanares
用地面积：605m²
建筑面积：55m²
总楼面面积：554m²
设计时间：2010
竣工时间：2011
造价：EUR 800,500
摄影师：©Antonio Arévalo(courtesy of the architect)

1 入口 2 装置/储藏室 3 等候室 4 接待处 5 行政办公室
6 警务办公室 7 报告办公室 8 健身房 9 衣帽间 10 警官办公室 11 镇议员办公室
12 董事会会议室 13 卫生间 14 民防办公室 15 休息室 16 武器库 17 监狱

1. entrance 2. installations/store 3. waiting room 4. reception 5. administration office
6. official office 7. reports offices 8. gymnasium 9. locker room 10. chief police office
11. town councilor office 12. boardroom 13. toilet 14. civil defense office 15. recess 16. weapon room 17. cell

一层 first floor

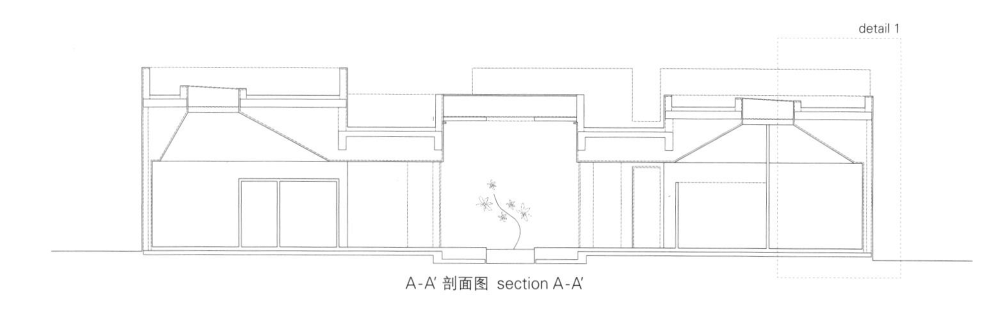

A-A' 剖面图 section A-A'

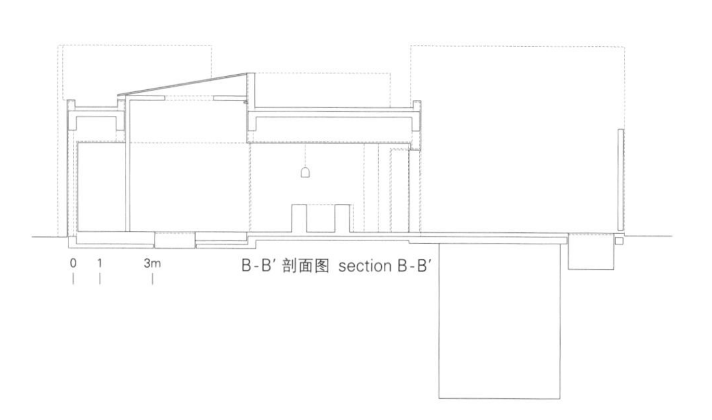

B-B' 剖面图 section B-B'

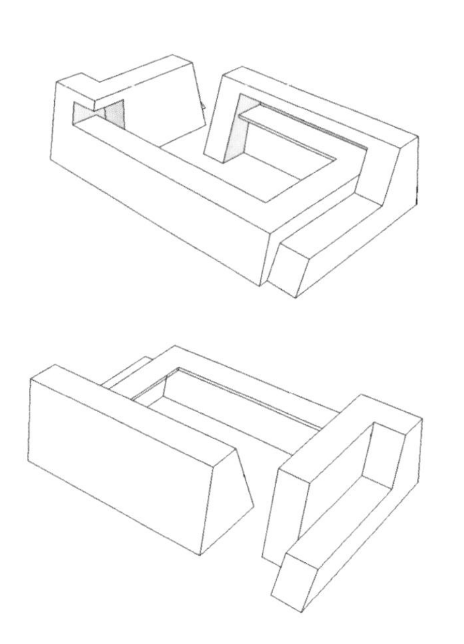

1. ceramic tile for facades anchored to a substructure
2. thermal clay block. dimensions 300x190x192mm
3. layer of asphalt
4. patching of slabs and beams with ceramic brick
5. layer of concrete of 10cm thickness
6. cement mortars of class M10
7. isolation consisting of a direct injection of polyurethane rigid foam
8. air chamber
9. structure profile galvanized steel 48mm wide, base studs(vertical elements) separated between them 600mm and channels(horizontal elements) fixed to the slab and wall
10. septum formed by self-supporting panel in plasterboard, painted and smooth
11. plinth of porcelain tile
12. expanded polystyrene of 3cm thickness
13. porcelain floor tile
14. mortar layer of 9cm thickness
15. under floor heating
16. isolation of 2cm thickness
17. layer of asphalt
18. reinforced concrete floor slab of 15cm thickness
19. polyethylene film
20. subslab gravel layer of 15cm thickness
21. floor insulation and drainage consisting of extruded polystyrene base layer and a high performance porous concrete
22. fiberglass reinforced plastic sheet
23. leveling layer of cement mortars of class M5
24. capping special piece consisting of the same tile as facade
25. removal translucent: polycarbonate supported on profiles
26. white lacquered aluminum L-profile bolted to the wall
27. suspended plasterboard ceiling

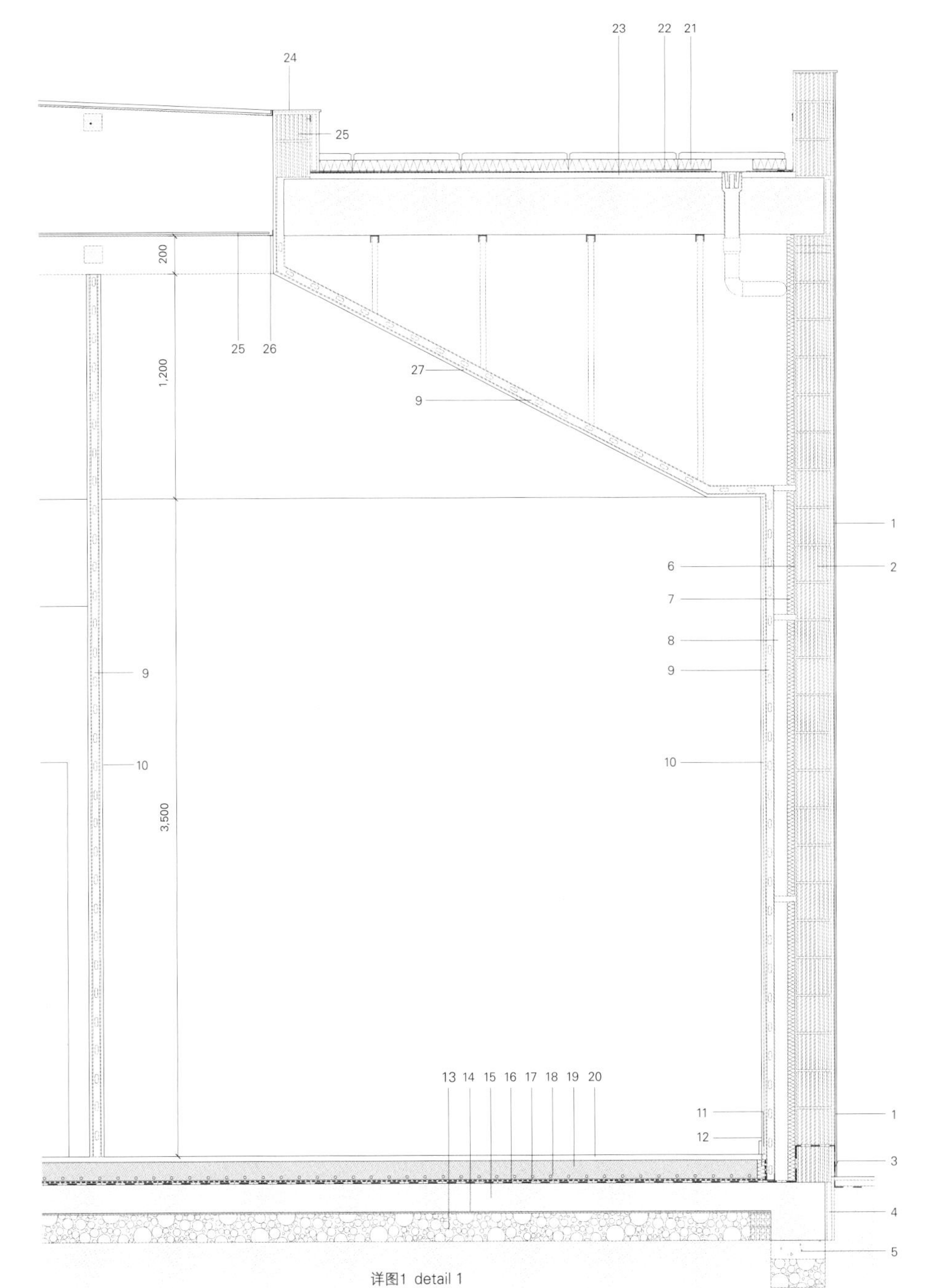

详图1 detail 1

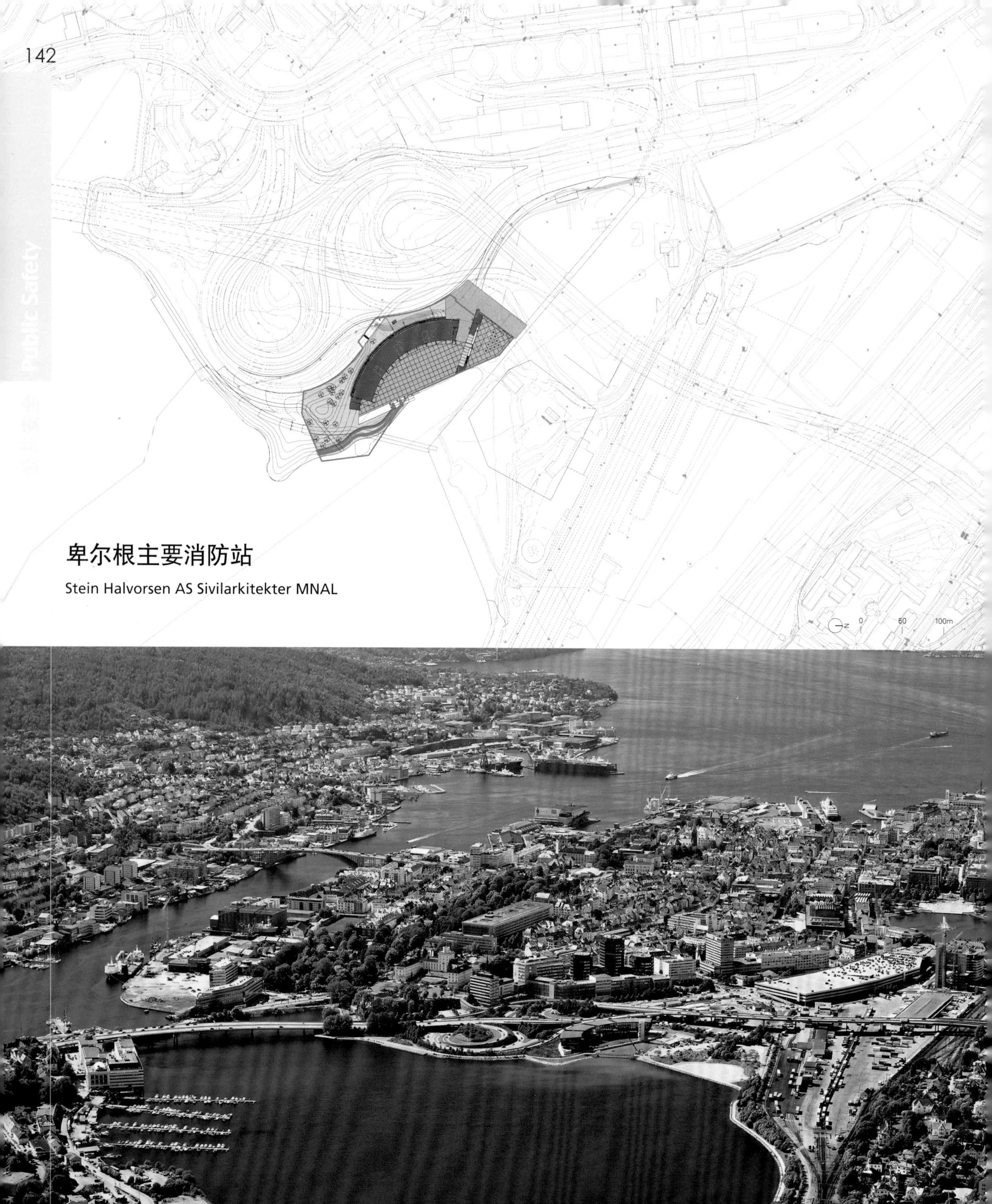

卑尔根主要消防站

Stein Halvorsen AS Sivilarkitekter MNAL

该消防站的设计有三个重要的基本要素：

－场地质量：该项目位于海岸线地带，在这里能够欣赏到城市山脉（乌尔瑞肯和Fløyen）的壮丽风景。

－占主导地位的拥堵的交通

－该建筑是本地区未来住宅区和城市开发项目的一部分。

该建筑呈弯曲状，外表设有一面可以遮挡住车来车往的密目网，在建筑前身形成了一个遮护空间。曲面是沿着场地的外部边缘设计的，这样能使建筑拥有尽可能宽敞的室内空间。该设施具有四大特色：基座、密目网、塔楼和桥梁。这种形式原理可从其外观、功能、结构和材料四个方面反映出来。

基座（一层和二层）包括海岸线、前院以及坚固的南端结构，南端逐渐与结构板和立柱融合在一起。该结构越往北越开阔，这种设计对该区域的未来住宅区以及城市开发项目起到了促进作用。

整个基座都是由现场浇筑的混凝土构成的，次要的结构构件是钢铁和铝材，从而形成了一种粗犷的外观效果。车库、工作室、紧急更衣室、技术室和储藏室都位于这较低的两层内。

密目网（三层和四层）是该设施最突出的元素。它悬挑于基座之上。其设计旨在使屋顶、西立面和介于二、三层之间的楼层平面能够成为一个整体。为了突出其外观，覆层选用了金属色的薄板。屋顶上的钢板铺有泡沫玻璃。金属色覆层在玻璃单元中依靠金属托架支撑。密目网内设有为应急人员准备的卧室和起居室以及行政办公室。密目网面向前院、大海和山脉展开。面朝前院的立面全部采用未经加工的落叶松木和玻璃打造而成。细长的垂直立柱有一种很清晰的韵律感，并突出了曲面的形状。这些柱子可作为立面的抗风支撑，并支撑着大型窗户。屋顶以及四层的天花板呈倾斜状，使人们能够欣赏到山脉或者更多的景色。二层的所有房间都设有直接通往阳台的通道，阳台设在密目网的下面，与整座建筑一样长。密目网的结构构件为钢结构和空心板。带有波纹板的钢梁支撑着屋顶，覆层除了玻璃之外，主要都是使用木材。目的是打造出一种经过精致加工的外观效果，以木材、玻璃和光亮的表面为特色。

塔楼位于基座之上，这种不对称的布置和前院有关，形成了一种张力。塔楼是该设施的统一体，标志着该设施在当地社区的重要性。塔楼是根据消防训练所需而设计的，比如楼梯井中的潜烟训练。阳台既是潜烟训练的休息处，也是救援练习的地方。

桥梁横跨于密目网之下，并与塔楼相连。它距离地面数英尺，在没有形成障碍的同时还在视觉上强调了城市和消防服务区之间的界线，并提供了很好的视觉联系。它在这个非常区域形成了一种清晰的特征：公众在不妨碍消防部门训练的情况下，可以从桥梁上俯瞰到他们的各种活动。桥梁内设有可以直达塔楼的演习室。桥梁同时也是消防部门停车场的屋顶。桥梁的主要结构构件是两排高大的钢桁架，健身室是一个嵌入桁架的盒状结构，该结构由浅色硬砖构成。之所以选用这种材料，是因为它能承受高温（200℃～300℃），而且保温性能优良。桥梁的覆层使用的是未经加工的落叶松木，以使整座桥看起来像是一根坚固的木梁。

Bergen Main Fire Station

There are three important factors fundamental to the design of the fire station:

- Site qualities: shoreline and magnificent views of the city's mountains, Ulriken and Fløyen
- Dominant and negative appearance of the traffic
- The building as a part of a future settlement and urban development in the area

Curved shape of the building with the dense screen against the traffic creates a sheltered space in front. The curve follows the outer edge of the site in order to keep this space as wide as possible. Facility includes four main features: the base, the screen, the tower and the bridge. This formal principle is reflected in the expression, functions, structures and materials.

The base (1st and 2nd floor) includes the shoreline, the front yard and the solid south end which gradually dissolves into structural slabs and pillars. The increasing openness of the structure to the north is an invitation to future settlement and urban development in the area.

The entire base is made of concrete, cast on site. Secondary structural elements are steel and aluminum. Goal has been to promote a rough look. Garages, workshops, emergency dressing rooms, technical rooms and storage rooms are located in these two lower floors.

The screen (3rd and 4th floor) is the most prominent element of

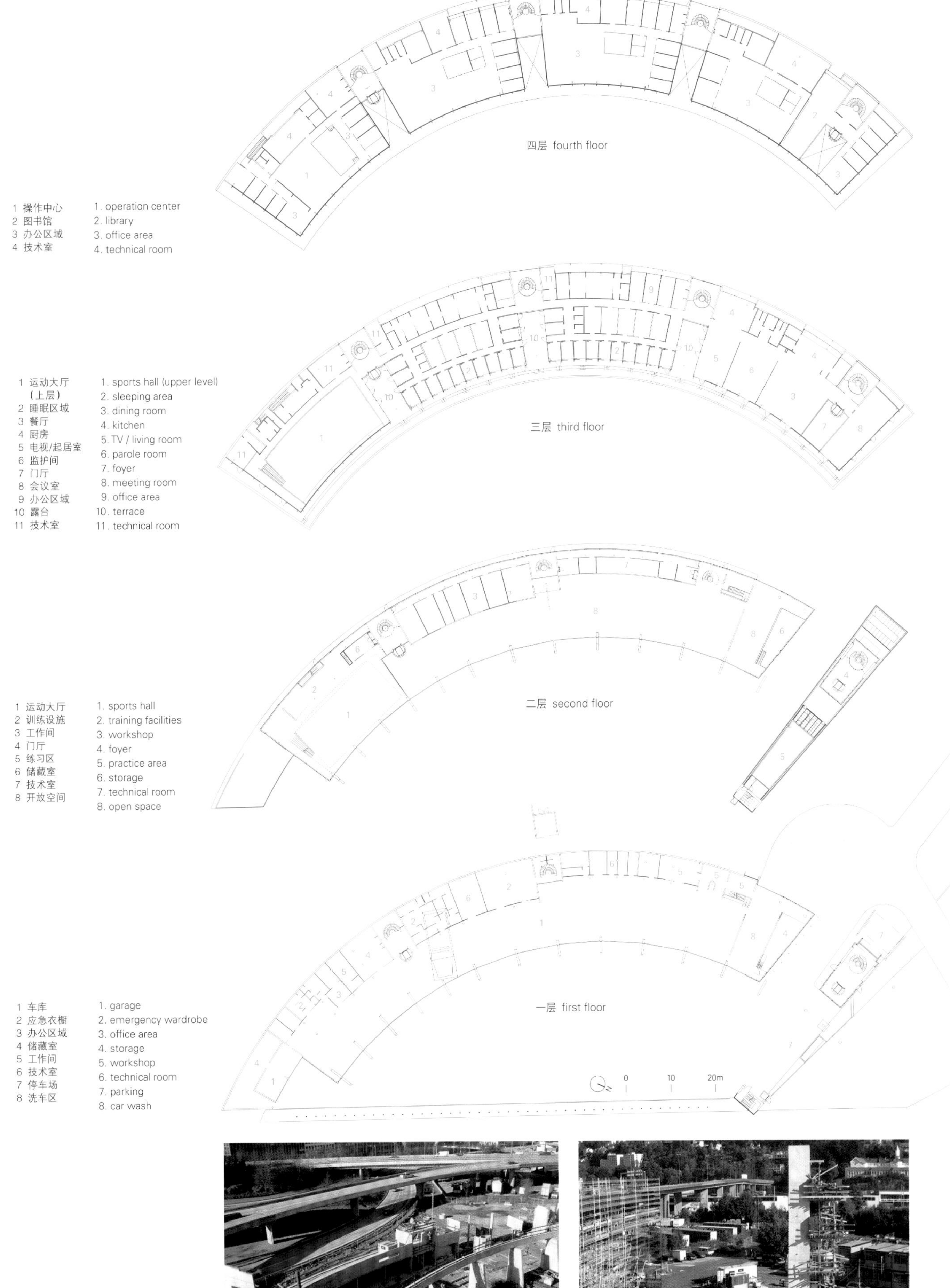

四层 fourth floor
1 操作中心 1. operation center
2 图书馆 2. library
3 办公区域 3. office area
4 技术室 4. technical room
三层 third floor
1 运动大厅（上层） 1. sports hall (upper level)
2 睡眠区域 2. sleeping area
3 餐厅 3. dining room
4 厨房 4. kitchen
5 电视/起居室 5. TV / living room
6 监护间 6. parole room
7 门厅 7. foyer
8 会议室 8. meeting room
9 办公区域 9. office area
10 露台 10. terrace
11 技术室 11. technical room
二层 second floor
1 运动大厅 1. sports hall
2 训练设施 2. training facilities
3 工作间 3. workshop
4 门厅 4. foyer
5 练习区 5. practice area
6 储藏室 6. storage
7 技术室 7. technical room
8 开放空间 8. open space
一层 first floor
1 车库 1. garage
2 应急衣橱 2. emergency wardrobe
3 办公区域 3. office area
4 储藏室 4. storage
5 工作间 5. workshop
6 技术室 6. technical room
7 停车场 7. parking
8 洗车区 8. car wash
0
10
20m

the facility. It rests on and cantilevers over the base. The screen is designed so that the roof, the west facade and the floor plane between 2nd and 3rd floor stand out as one element. In order to emphasize the shape, natural copper plates are chosen for cladding. The roof is compact with cellular glass on steel plates. Copper cladding is attached to the metal brackets in the glass cells. The screen contains bedrooms and living room for the contingency crew and administrative functions. The screen opens towards the front yard, the water and the mountains. The facade facing front yard is entirely made of untreated larch and glass. Slender vertical pillars clarify the rhythm and emphasize the curved shape. The pillars function as wind bracing for the facade and bearing for the large windows. The roof and the ceiling of 4th floor are tilted to open up towards the view of the mountains even more. On the 2nd floor there is direct access to balconies from all rooms. The balcony is lowered into the screen in the whole length of the building. The structural elements of the screen are steel structure and hollow core slabs. The steel beams with corrugated sheeting support the roof, while cladding in addition to glass is mainly wood. The goal has been to promote a refined, processed expression, characterized by wood, glass and bright surfaces.

The tower rises from the base. The asymmetric placement in relation to the front yard creates tension, while the tower serves as a unifying object in the facility and signals facility's importance for the community in general. The design of the tower is due to fire training requirements, such as smoke diving in the stairwell. Balconies are used both to rest during smoke diving exercise, and to practice rescuing people.

The bridge spans under the screen and connects with the tower. It is lifted a few feet of the ground to visually highlight division between the city and the fire service area without forming a barrier and provide for good visual connection. It signals a clear distinction to an area that requires special considerations; the general public is able to overlook activities of the fire department from the bridge without being in the way. The bridge contains rehearsal rooms with direct connection to the tower. It also serves as a roof over the fire department parking lot. The main structural elements of the bridge are two high steel trusses. Exercise room is a box of light clinker brick fitted in the truss. The material is chosen because of its ability to withstand the high temperatures (200-300ºc) and its insulating properties. Cladding of the bridge is untreated larch to give the whole bridge appearance of a strong wooden beam. Stein Halvorsen AS Sivilarkitekter MNAL

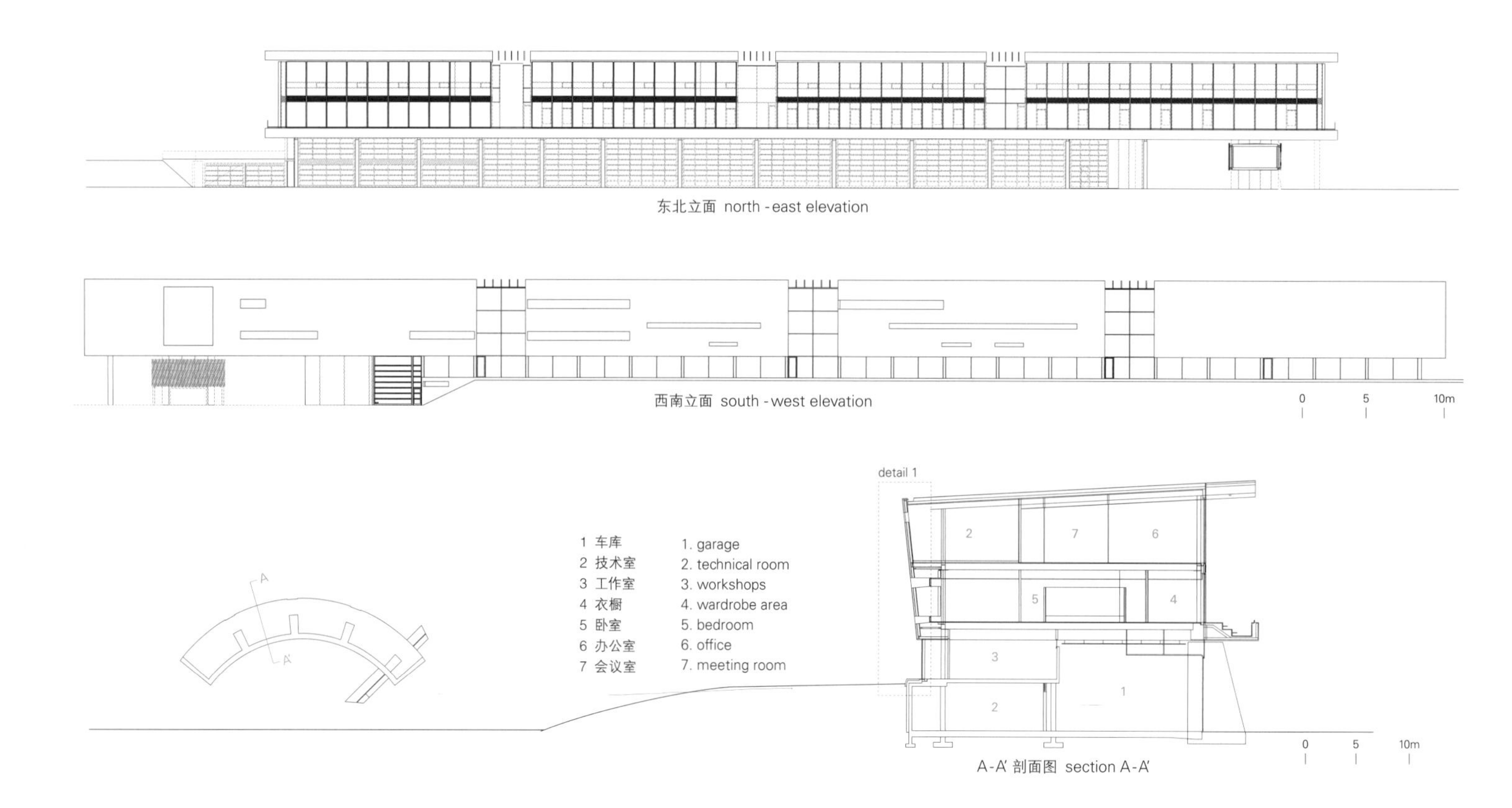

东北立面 north-east elevation

西南立面 south-west elevation

A-A' 剖面图 section A-A'

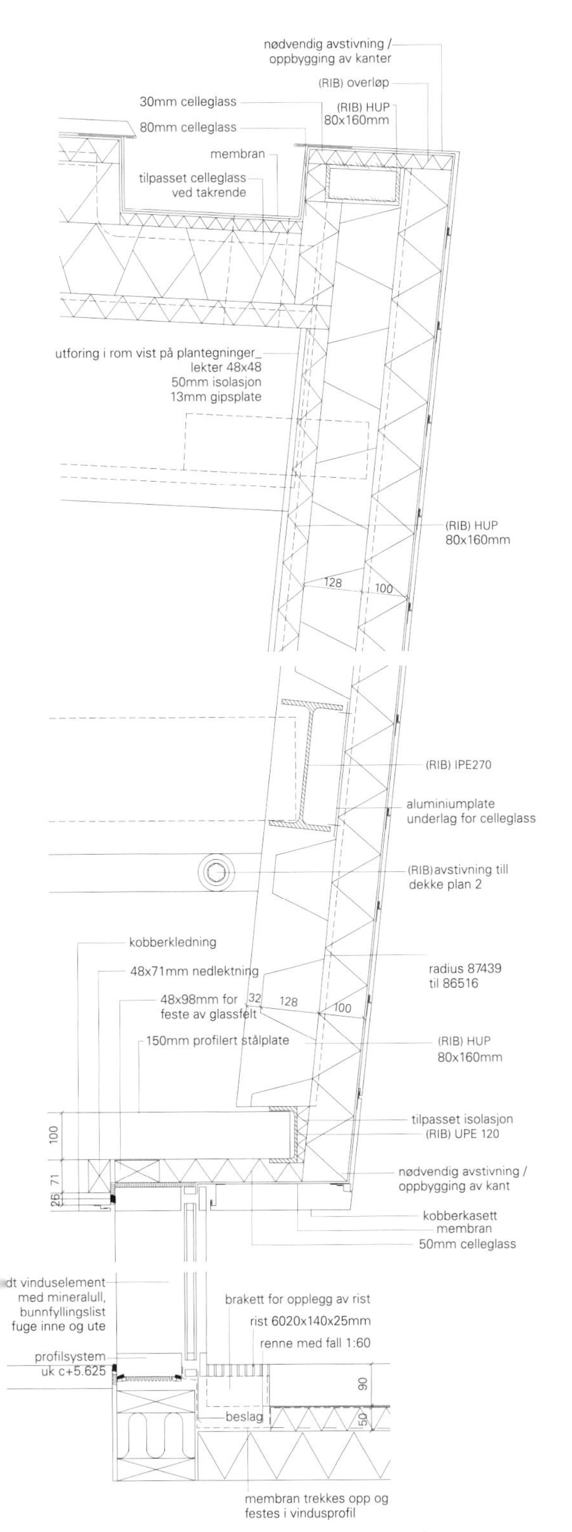

详图1 detail 1

项目名称：Bergen Main Fire Station
地点：Store Lungegaardsvann, Bergen, Norway
建筑师：Stein Halvorsen AS Sivilarkitekter MNAL
合作者：
Magnus Rynning-Tønnesen sivilarkitekt MNAL (project leader),
Svein Tore Haugen Sivilarkitekt MNAL,
Camilla Kolbenstvedt Sivilarkitekt MNAL,
Thomas Lykke Nielsen Arkitekt MAA,
Bjørner Bolle Sivilarkitekt MNAL
建筑商：BKB – Bergen Kommunale Bygg
室内设计：Stein Halvorsen AS Sivilarkitekter MNAL
景观建筑师：Grindaker AS
合作者：Skanska Norge AS
结构工程师：AS Frederujseb
HVAC/卫生：Rambøll Norge AS
电力工程师：AS Rasmussen & Strand
防火顾问：Rambøll Norge AS
交通顾问：Norconsult Bergen AS
总楼面面积：8,000m^2
竣工时间：2007 造价：180 mill NOK
摄影师：©Kim Müller(courtesy of the architect)

岩石中的Margreid消防站

Bergmeister Wolf Architekten

新落成的Margreid市志愿者消防站坐落在一块岩石表面上，这里位于Wine大街。建筑师在岩石内部开凿出三个巨型洞穴，彼此之间由一条横向切口相连接。一堵被涂成黑色的混凝土墙挺立在岩石前方，与山体的坡度相同。三个洞穴的入口与这堵混凝土墙相衔接。

这堵墙壁是消防站的主要建筑元素，同时也可以阻挡住滚落的石块。因此，该设计既巧妙地处理了地形状况，又满足了使用需求。建筑师之所以选择混凝土作为墙体材料是因为它耐用、坚固、强大。其黑色的外观由于涂上了榉木炭灰，看上去好似一块烧焦的木头。

三个车库穿透了这堵墙壁，其中两个车库探了出来，构成了建筑入口，外面覆有涂成黑色的钢材。这些车库采用的是玻璃折叠门，因此，在外面就能看见红色的消防车。在办公室和行政侧翼所处的区域，一个悬挑式玻璃立方体穿透了墙壁，在立方体里可以观赏到四面八方的景色，同时，这一设计还将光线引到了室内。

洞穴内部的特色在于材料简单：木材、玻璃以及钢材的运用手法细腻，与洞穴粗糙的抹灰表面形成对比。

在可持续性方面，这座消防站与其他项目有两点区别：建筑选址和整体能源理念。这座建筑本可以像大多数房屋一样建造在一片正常的场地上，但由于高山地区土地稀少，因此社区决定把第一座消防站建到岩石里。从而使珍贵的土地节省下来，可以为农业服务。这意味着该项目在节约资源方面做出了积极贡献。为了减少运营成本，促进环境保护，Margreid社区希望建造一座节能建筑。这种想法同样促使人们决定将建筑设在岩石中，只留玻璃入口面对外界（冬季温度为−10℃）。建筑其余部分都置于岩石内部（冬季温度为12℃）。这样一来，供暖成本降低了，能源也就被节省了下来。另外，通过动态模拟热量流动的方法，建筑师可以挑选出必须要经过保温处理的岩石部位。

结果证明，只有站内的行政侧翼需要覆盖保温层，车库则可以利用周围岩石的天然温度加热。其余的玻璃表面采用的则是三层吸热玻璃。供暖（40kWh/m²）和烧水所需的其余能源则由生态球丸燃料加热系统提供。

In the Rock, Margreid Fire Brigade

A rock face is the location of the new volunteer fire brigade of Margreid on the Wine street. Three big caverns are drilled into the rock and interlinked with a cross cut. A black pigmented concrete wall stands in front of the rock, with the same inclination as the mountain. The three caverns dock on to this concrete wall.

The wall is the main architectural element of the fire brigade and at the same time a protection against down falling stones. Therefore resulted a sensitive handling with the topography and the requirements of use. Architects chose concrete as the material for the wall: durable, strong and powerful. The dark colour is achieved by the application of beech coal dust and should bear resemblance to burnt wood.

Three corpuses penetrate this wall. Two garages break through

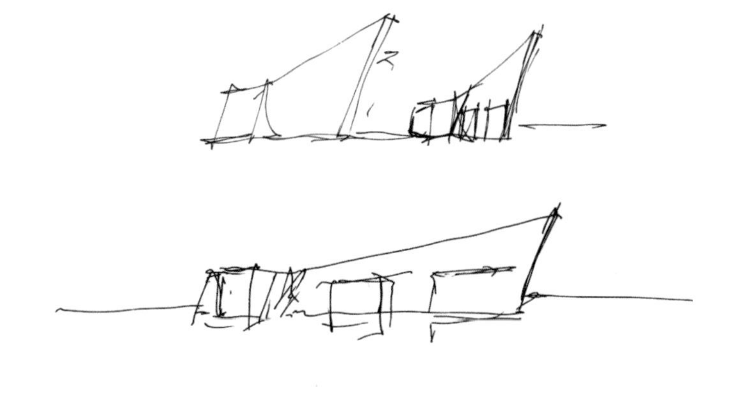

forming portals which are covered with black coat steel. These garages are closed with glass folding gates which allow the red fire engines to be seen from the outside. In the area of the office and administration wing the wall is broken through by a cantilevered pending glass cube which gives free sight in all directions and brings light into the inside.

The interior of the caverns is characterized by simple materials: wood, glass and steel are used subtly and contrast the cavern with its harsh plastered surface.

In terms of sustainability the fire brigade differentiates itself in two points from other projects: the positioning of the building and the overall energy concept. The building could have been placed on a normal lot as most of the houses, but because of the rare grounds in the alpine context, the community decided to build the fire station into the rock. Therefore valuable ground has been saved and can be used for agriculture. This means an active contribution to save resources. In order to reduce running costs and contribute to environment protection the community of margreid wanted to build an energy-efficient building. This also encouraged the decision to place the building into the rock where only the glass entrances face the outside (-10°C in winter). The rest of the building lays inside the rock (+12°C in winter). As a result heating costs is reduced and therefore energy is saved. In addition, a dynamic simulation of the heat flow allowed to select the parts of rock which had to be insulated.

Consequently only the administration of the station had to be covered with thermal insulation while the garages could be heated with the natural temperature of the surrounding rock. The remaining glass surfaces are carried out in triple heat absorbing glass. The residual energy for heating (40kWh/m²) and hot water is covered with an ecological pellets heater system.

Bergmeister Wolf Architekten

1 入口
2 会议室
3 厨房
4 浴室
5 更衣室
6 陈列室
7 储存室
8 车库
9 控制室
1. entrance
2. conference room
3. kitchen
4. bathroom
5. changing rooms
6. showroom
7. storage
8. garage
9. control room
一层 first floor
0
5
10m
A-A' 剖面图 section A-A'
B-B' 剖面图 section B-B'
东立面 east elevation

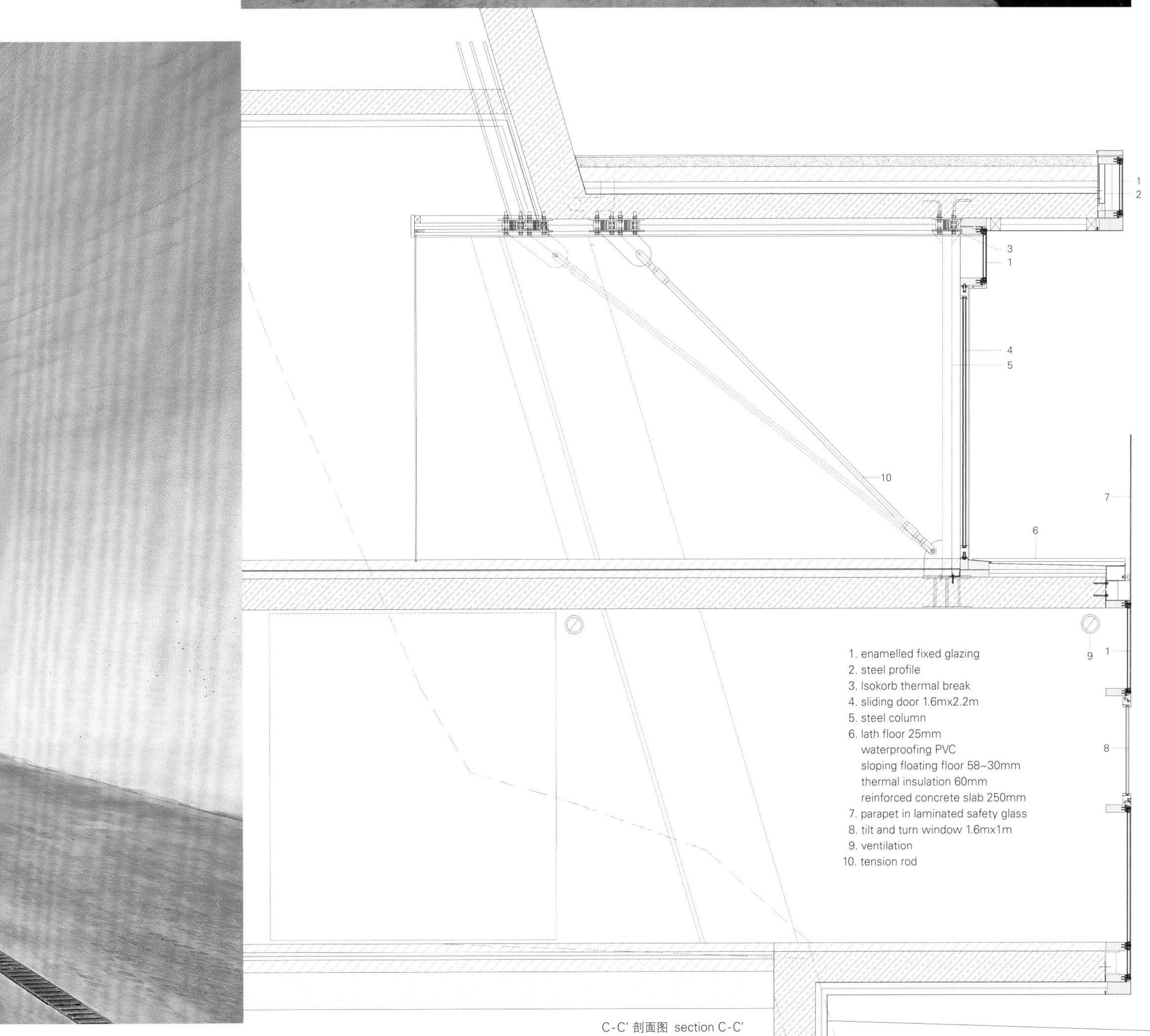

C-C′ 剖面图 section C-C′

项目名称：In the rock, Margreid Fire brigade
地点：Margreid, Italy
建筑师：Gerd Bergmeister, Michaela Wolf
合作商：Christian Ghedina, Jürgen Prosch
隧道/静力：Planteam Gmbh
甲方：Gemeinde Margreid an der Weinstrasse
用地面积：2,712m²
总楼面面积：690m², 259.57m² above ground and 430.43m² underground
用途：hall of fire department equipment Margreid
设计时间：2009 施工时间：2009—2010
摄影师：courtesy of the architect-p.148, p.150~151, p.155
©Jürgen Eheim(courtesy of the architect)-p.153, p.154, p.156, p.157[top]
©Günter Richard Wett(courtesy of the architect)-p.157[bottom]

斯蒂芬·霍尔最近设计博物馆的新情境主义

A New Contextualism on Steven Holl's Recent Museums

世界是水平的。全球化把这个世界变成了一个水平的运动场，人们可以享有更多平等的机会，地理位置变得越发不重要。不过，副作用也随之而来，那就是我们创造的环境越发地相似和重复。雷姆·库哈斯设计过很多飞机场和酒店房间，他的这些设计给人一种雷同感。还有另一个极端，一些建筑师努力地效仿当地文化，这必然使他们的设计作品给人一种照本宣科的肤浅之感。在全世界追逐身份个性的潮流中，能够在"环境至上"和"一味雷同"两个阵营中找到平衡点的建筑师是屈指可数的，美国建筑师斯蒂芬·霍尔就是其中的一位。他自身并没有一个明确的"风格"，但是通过独特的表达方式，他不断巧妙地回应了"环境"所提出的难题，这在他的博物馆工程中表现得尤为突出。在赫尔辛基，当代艺术博物馆怀抱着这个城市的建筑遗迹和自然风光，呈现了引人入胜的交织形态和空间。在堪萨斯城，五座相连的玻璃建筑体量与古老的石造建筑——纳尔逊艺术博物馆相得益彰，几何学与光学的诗意运用给美术馆以独特的品质，并为现代艺术营造了中规中矩但不失活力的空间。位于挪威的克努特·汉姆生中心采用了Hejdukian"将建筑作为主体"的概

The world is flat. Globalization has turned it into a level playing field where people enjoy more equal opportunity and their geographic locations become less relevant. The side effect, unfortunately, is that our built environment appears to be more and more equalized and repetitive. Rem Koolhaas' intense experience in airports and hotel rooms brings the generic to our attention. On the other end of the spectrum, some architects try too hard to mimic local culture and their work becomes inevitably literal and kitschy. In this global struggle of identity, American architect Steven Holl is one of the few who can still remain in balance between the "Fuck-Context" camp and the "Cheesy-Copycat" operations. He does not have a definite "style" per se but he has repeatedly answered the question of "context" skillfully with his distinct forms and narratives, especially in his museum projects. In Helsinki, the Kiasma Museum of Contemporary Art embraces the city's architectural legacy and natural landscape, resulting in interesting intertwining shapes and spaces. In Kansas City, five interconnected glass volumes co-exist with the old masonry building of Nelson-Atkins Museum of Art in complementary contrast. The poetic play with geometry and light gives the galleries different qualities and creates neutral yet dynamic spaces for contemporary art. The Knut Hamsun Center

芬兰赫尔辛基当代艺术博物馆，建于1998年
Kiasma Museum of Contemporary Art in Helsinki, Finland, 1998

照片提供：©Ernst Furuhatt
挪威克努特·汉姆生中心，建于2009年
Knut Hamsun Center in Hamarøy, Norway, 2009

照片提供：©Marc Teer
美国堪萨斯城纳尔森·阿特金斯艺术博物馆，建于1999年
Nelson-Atkins Museum of Art in Kansas City, USA, 1999

念，采用了当地的材料和建筑传统，例如染黑木和长草皮屋顶。

从最近刚刚完工的两座博物馆——法国比亚里茨的海洋与冲浪之城和中国南京的四方艺术博物院中，我们可以再次看到斯蒂芬·霍尔对环境的出色敏感度，以及他在给建筑注入新特征的同时突出建筑的环境特点的能力。

风景

两座博物馆都坐落在大自然的环境中，它们与周围风景的关系是至关重要的。一个沉浸于海边的美景，另一个则以山峦为背景，这两个工程以不同的策略回应自然。

比亚里茨是法国一个美丽的海滨小镇，有着大西洋上海天相接的美景。海洋与冲浪之城的右边是Plage la Milady海滩，位于市中心之南的Ilbarritz区域。在这深邃的蓝色面前，霍尔选择减小建筑体积。设计以"在蓝天下，在大海中"为名，建筑被压缩在弧形的屋顶下，弧形结构形成了一个上层的凹面公共广场和展览室上方生动的凸面天花板。

这种类地形设计成功地融合了风景和建筑，创造了一系列戏剧性的地形变化。一段缓坡楼梯从建筑前的大街直通广场一层，广场的两侧是卷曲的，能够领略到远处海天相接的美景。在其中一侧，广场向上倾斜至一个平台，它是欣赏美丽日落的最佳地点之一。在另一侧，广场向下由斜坡通向平地，能量流通过一个草地公园进入广阔的海洋。

在广场上，只能看见建筑的两个"玻璃岩石"，它们的功能分别是餐厅和冲浪凉亭，外观与比亚里茨海岸上随处可见的巨大岩石非常相似。不过，Okalux玻璃的材质赋予这些"岩石"可反光的柔和质感，这种朦胧感与岩石的粗糙坚硬形成了反差。这种抽象的类比法充分展现了霍尔的设计可以使建筑与周围环境产生敏锐和诗意的联系，而并非是生搬硬套的联系。

欧亚大陆的另一端，在中国南京西北的一处度假胜地——珍珠泉，四方艺术博物馆坐卧于浓密的树林里，它有老山国家森林公园山峦叠翠的美丽景色作为背景，以及前面的佛手湖作为点缀。霍尔选择了一种强烈的表达方式，但是仍不失和谐。建筑外部一大片黑色的墙壁与周围的

in Norway adopts the Hejdukian concept of "building as a body" and applies local materials and building traditions such as stained black wood and long grass sod roofs.

In two recently completed museums – the City of Ocean and Surf in Biarritz, France and Sifang Art Museum in Nanjing, China, Holl demonstrates once again his unique sensibility of context and ability to reinforce the locale of the projects while injecting new characters into the place.

Landscape

The two museums are both situated in a natural context, where the relationship to the landscape is utterly important. One immersed in the landscape next to the ocean while the other standing out from the mountain background, the two projects take on different strategies in response to nature.

Biarritz is a beautiful French beach town with fantastic views towards the Atlantic horizon. The site for the City of Ocean and Surf is right next to the Plage la Milady in the Ilbarritz area south of the city center. In front of the grand blue, Holl chooses to reduce the building volume. Dubbed "under the sky / under the sea", the design concept is to press the building in the ground beneath a curved roof that forms a concave public plaza above and a dramatic convex ceiling in the exhibition spaces below.

This quasi-landform successfully renders the fusion of landscape and architecture, and creates a theatrical sequence of topographic changes. A gentle stairway leads up to the plaza level from the street front of the building. Both sides of the plaza curve up, flanking the ocean view towards the horizon in the distance. On one side, the plaza rises to a terrace – one of the best spots to watch an amazing sunset. On the other side, the plaza slopes down onto the ground, leading the energy flow through a grassed park into the vast ocean.

The only visible parts of the building on the plaza are two "glass rocks" functioning as a restaurant and a surfers' kiosk. They have a strong physical resemblance to the giant boulders on the seashore, which are pervasive characters along Biarritz's coastline. On the other hand, the Okalux glass gives the "rocks" a diffused and soft materiality, an ambiguity that is contrary to the roughness of stone. This abstract analogy is a perfect illustration of Holl's sensible and poetic link to the context – it is there, but not literally.

On the other side of the Eurasia continent, Sifang Art Museum is located deep in the woods of Pearl Spring, a resort northwest of Nanjing, China. The site is strikingly picturesque with mountains

景色融合，其中一座白色的建筑在树林间挺立，在郁郁葱葱的山林里异常地显眼。在活动空间和阳台上，通过建筑内部蜿蜒曲折的通道，空间感被彻底打开，古老都城的遥远景象尽收眼底。将视觉与环境结合是设计界的传统美德，但它已经被很多现代建筑师遗忘了。

敞开的白色建筑看似漂浮在空中，这是因为其下的支撑结构很小，几乎是看不到的。这个高架桁架管结构只依靠三个"腿"来保持平衡，像一个三脚架，由楼梯、电梯核心筒和薄薄的剪力墙组成。这看似脆弱的支撑体，却足以撑起整个建筑，原因是外墙是由轻质但耐用的聚碳酸酯材料制成的，这种材料提供了柔软、朦胧的品质，与比亚里茨的海洋与冲浪之城的玻璃岩石有着相似的特点。

文化

对于法国和中国来说，霍尔都算是外国人。他相信"未来植根于历史的土壤"，试图理解更加广阔的文化背景，进而从多种多样的传统和历史遗迹中获得灵感。在形形色色的中国文化中，霍尔选取了中国传统水墨画的视角转换和空间分层的理念，运用到四方艺术博物馆的设计里。文艺复兴时期的艺术家们在作画时惯于选取单一的消失点，而中国的画家们却否定了这个想法，进而创造了具有平行视角的山水画画法。"无限焦距"创造了多样性空间和非线性场景，使得深度感不依赖于视觉结构。按照这独特的中国式技巧，霍尔将建筑的下部设计为一系列移动的墙壁，墙上满是缺口和转弯，形成了视角平行的空间。空间的复杂性在上方浮动的美术馆里延续，偶尔会有朝向不同的裂缝窗口，给游客带来更加丰富的体验。

说得更具体一些，霍尔是将文化背景与材料、细节的选择联系在了一起。那一大片黑色墙壁是竹子结构与混凝土组合而成的，这种特殊的建筑技巧是专门为这个工程设计的，方法是事先让竹子生长在这里，使竹子形成混凝土的独特支撑结构。建筑的颜色仅限于黑、灰和白三种，这是为了将建筑与中国传统水墨画的意境联系在一起。所有这些精妙的处理方式都充分反映了霍尔对当地文化的高雅诠释。

在比亚里茨，博物馆的设计形态同样紧紧联系了这个海岸城市生机勃勃的冲浪文化，弯曲的轮廓代表着大海的滚滚波涛，同时让人模糊地联想到了滑冰坡道。虽然坡道被铺满了鹅卵石无法进行滑板运动，但是

of Laoshan National Park as background and the waters of Foshou Lake in the front. Here, Holl opts for a stronger yet still harmonious presence. Out from a "field" of black walls that blend more into the landscape, a white "figure" rises up and hovers above the trees, distinctly visible from its lush hilly surroundings. Through the winding passage in the building, the spatial experience unwraps and peaks at the top where on the event space and its balcony people enjoy a distant view to the ancient capital city. This effort to make visual engagement with the context is a traditional virtue that is almost forgotten by many contemporary architects.

The unraveling white figure seems to be floating in the air – the structural support is so minimal that it is almost invisible. This elevated trussed tube structure is balanced only on three "legs" like a tripod: a staircase, an elevator core, and a thin shear wall. One of the reasons why this is sufficient is that the cladding is a lightweight but durable polycarbonate material. It also gives the volume a soft and ambiguous quality similar to the glass boulders on the plaza of the City of Ocean and Surf in Biarritz.

Culture

Both in France and China, Holl is a foreigner. Believing that "the future grows from the soil of history", he tries to understand the broader cultural context and his design takes inspiration from various traditions and heritages. From the many aspects of Chinese culture, Holl selects the shifting viewpoints and layering of space in traditional Chinese painting as his reference point for Sifang Art Museum. While the Renaissance artists identified one single vanishing point in their paintings, the Chinese rejected that idea and continued to produce landscapes with parallel perspectives. The "infinite focal length" creates multiplicity of space and nonlinear scenes where the sense of depth does not rely on visual construction. With this special Chinese technique in mind, Holl shapes the lower part of the building with a series of shifting walls full of cuts and turns to form parallel perspective spaces. The spatial complexity continues above in the floating galleries, where occasional slit windows point at different directions and further enrich the visitor's experience.

In a more specific way, Holl relates to the cultural context with the choice of materials and details. The walls in the "field" are bamboo-formed concrete – a special technique developed for the project to utilize the bamboo previously growing on the site as formwork to achieve a unique texture of the concrete. The decision to limit the color palette to only black, grey, and white also connects the building to the poetic nature of traditional Chinese ink paintings. All these subtle treatments are perfect evidences of Holl's tasteful interpretation of local culture.

In Biarritz, the form of the building captures the strong and lively surfing culture of the coastal city. The curved profile resembles the waves of the ocean, and at the same time evokes a loose read-

照片提供：SHA Architects (©Iwan Baan)

在海洋与冲浪之城，人们在云状池中娱乐
City of Ocean and Surf, people playing in the cloud pool

照片提供：SHA Architects

南京四方艺术博物馆低矮的墙壁群
Nanjing Sifang Art Museum, the lower field of walls

照片提供：SHA Architects(©Li Hu)

通过落地窗遥望南京市中心，欣赏周围美景
The big window to the Nanjing city center and the surrounding landscape

广场仍然是一个充满活力的公共空间。在这里，人们奔跑、攀登、跳跃，好像这戏剧性的地势将人类隐藏的"猴子的一面"释放了出来。霍尔力图建造一个真正的滑冰池供冲浪运动员做实验。这种云状的凹陷为各种各样可以想象和不可想象的活动提供了舞台。

内容

在路边的一侧，海洋与冲浪之城高高的曲线形拐角下方，是展览馆的入口大厅。陡峭的斜面天花板强烈提示着人们空间将大幅下降到处于半地下的展览室。然而，抱着满满的期待进入展览室内部的人们会变得十分失望，虽然有高科技设备，但是展示设计看起来粗糙无力，与建筑设计毫无关系。

建造海洋与冲浪之城的目的是为了引起人们对于海洋问题的关注，促进冲浪与海洋的科学探索，以及在休闲娱乐、科学、生态方面发挥作用。最初，博物馆的董事会决定分别聘请展览设计师，这样信息共享和设计协调就变得非常困难。展览设计师Already Made跟随着"寓教于乐"的风尚，并效仿迪士尼奇特的幻想世界，把原应冷静和诗意的内部设计成了一个游乐园。

这让我联想到了丹尼尔·里伯斯金设计的位于柏林的犹太博物馆，该馆混乱的展览安排与建筑风格完全不协调。相似的案例还有扎哈·哈迪德设计的位于沃尔夫斯堡的菲诺科学中心，现代博物馆的建筑设计和展览设计不能互相合作真是令人遗憾。由于建筑师的专业知识有限，他们忽略了内容，只注重一般的外壳，丢弃了项目背景。拥有像纳尔逊·阿特金斯艺术博物馆那样整体质量的博物馆现在已经寥寥无几了。

南京四方艺术博物馆现在虽然完工，但却没有定期向公众开放。设计之初，展览设计就缺少一个经营者或者一个可靠的项目。该展览属于中国国际实用建筑展览活动组织的一项大型计划的一部分，由矶崎新负责，该组织成员还有王澍、SANAA、戴维·埃德加耶、Mansilla+Tuñón等建筑界精英。中国国际实用建筑展览项目是以"用营造环境的形式来颂扬艺术、文化和大自然"为主题建立起来的。霍尔自由地设想了一个为艺术和建筑而开辟的空间。在去年博物馆即将建成的时候，四方博物馆得到正式命名，并成为组织的一个新分部。这个建筑高雅地反映了自然文化背景，我们希望这个新的机构及其组织者都能够珍惜这个独特的家。

ing of skate ramps. Although paved in cobblestones as an anti-skateboarding measure, the plaza is still a truly active public space. People run, climb, and jump, as if the "monkey side" of Homo sapiens were released by this dramatic topography. At the southwest corner, Holl has managed to put in a real skate pool for the surfers' experiments. The cloud shape indentation becomes an instrument for all sorts of unimaginable and imaginative activities.

Content

Under the high corner of the curve on the roadside of the City of Ocean and Surf is the entry lobby to the exhibitions. The steep sloping ceiling intensifies the spatial indication of diving down to the semi-underground exhibition space. But after all the anticipation, the exhibition inside is quite a letdown – even with the high-tech equipments, the display design seems rough and weak, and it has nothing to do with the architecture.

The intent of the City of Ocean and Surf is to raise awareness of oceanic issues and explore scientific aspects of surf and sea, and their role upon our leisure, science, and ecology. From the very beginning, the museum board decided to hire exhibition designers on a separate contract, which has made information sharing and design coordination extremely difficult. The exhibition designer Already Made follows the trend of "edutainment" and the fancy imagery of Disney, and their design turns a calm and poetic interior into an amusement park.

This reminds me of Daniel Libeskind's Jewish Museum in Berlin, where the chaotic exhibition arrangements are completely discordant to the architectural space. Similar disconnection happens in Zaha Hadid's Phaeno Science Center in Wolfsburg. It is unfortunate that architecture and exhibition design seldom go hand in hand with each other in contemporary museums. The limited influence of the profession has forced many architects to either ignore the content or just go for the generic white box. The program context is lost. The opportunity to achieve the holistic quality in the Nelson-Atkins Museum of Art is rather rare these days.

In Nanjing, Sifang Art Museum is now completed but not yet open regularly to the public. The design started without an operator or a solid program for the type of exhibitions, as part of a larger plan curated by Arata Isozaki for the China International Practical Exhibition of Architecture (CIPEA) that featured other stars such as Wang Shu, SANAA, David Adjaye, and Mansilla+Tuñón. The whole CIPEA project was organized under the loose theme of "celebrating arts, culture, and nature in the form of built environment". Holl took the liberty to envision a space for art and architecture. It was only at the very end of construction last year when Sifang Art Museum was officially named the building as their new branch. We can only hope that this young institution and its curators cherish their unique home that gracefully reflects the landscape and the cultural context. Human Wu

海洋与冲浪之城

海洋与冲浪之城是一座博物馆，它探索冲浪和海洋，也探索这二者在我们的休闲生活、科学和生态领域的作用。斯蒂芬·霍尔建筑师事务所和Solange Fabiao联合设计的项目方案赢得了一场国际竞赛，与他们同时参加竞赛的建筑事务所还有：Enric Miralles、Benedetta Tagliabue、Brochet Lajus Pueyo、Bernard Tschumi 和Jean-Michel Willmotte建筑事务所。

建筑的外形源自于"在蓝天下"和"在大海中"的空间理念。"在蓝天下"理念形成的凹状外观是外部空间——"Place de l' Océan"——的主要特色。凸面的结构天花板形成了"在海洋中"这一展览空间。人们一进入口就能感受到建筑的空间质量，并且在他们途经建筑的动态曲线表面（由于变幻的图像和光线而变得富有生气）时，可以通过入口处的大厅和坡道视线开阔地俯瞰展区。

建筑理念和地形的精确融合给予了建筑一种独特的外观。在朝向海洋的一面，该建筑的凹状平台越过景观延伸至大海。景观混合了田野和当地植被，在边缘处微微凹陷，它是建筑的延伸，将举办与博物馆设施相结合的盛事和日常活动。

两块"玻璃岩石"容纳了餐厅和供冲浪者使用的凉亭，它们使室外中心广场变得有活力，并似乎和远处海滩上的两块岩石联系起来。通过主入口大厅人们可以到达"玻璃岩石"，主入口大厅将街道、咖啡厅和供冲浪者使用的凉亭连在一起，但人们也可以通过向公众开放的主要聚集空间——广场，独立地进入。

博物馆的储藏室位于展览空间的中间层，可以由此直接进入入口大厅和礼堂。氛围较亲切的餐厅和抬升的户外露台在博物馆的顶层，并给人们提供开阔的海景。

在建筑的西南角有一个溜冰池，可供冲浪者在广场和下面的开放式回廊上休闲时使用；它还与礼堂和博物馆内部的展览空间相联系。这处带顶的空间为室外交流、聚会和活动提供了遮护。

建筑的外面是白色的变形混凝土饰面，这种混凝土由来自法国南部的骨料制成。广场上采用的材料是葡萄牙鹅卵石、草以及天然植被铺砌的，极富变化性。保温玻璃构件和透明的酸蚀层相结合，创造了一种视觉上的动感，也提高了内部的舒适度。主要空间的内部涂有白色石膏，而地板是木质的，下面设有缆线装置。

City of Ocean and Surf

The City of Ocean and Surf is a museum that explores both surf and sea and their role upon our leisure, science and ecology. The design by Steven Holl Architects in collaboration with Solange Fabiao is the winning scheme from an international competition that includes the offices of Enric Miralles, Benedetta Tagliabue, Brochet Lajus Pueyo, Bernard Tschumi and Jean-Michel Willmotte. The building form derives from the spatial concept "under the sky" and "under the sea". A concave "under the sky" shape forms the main character of the exterior space, the "Place de l'Océan". The convex structural ceiling forms the "under the sea" exhibition spaces. The building's spatial qualities are experienced already at the entrance where the lobby and ramps give a broad aerial view of the exhibition areas, as they pass along the dynamic curved surface that is animated by moving image and light.

The precise integration of concept and topography gives the building a unique profile. Towards the ocean, the concave form

of the building plaza is extended through the landscape. With slightly cupped edges, the landscape, a mix of field and local vegetation, is a continuation of the building and will host festivals and daily events that are integrated with the museum facilities.

Two "glass boulders", which contain the restaurant and the surfer's kiosk, activate the central outdoor plaza and connect analogically to the two great boulders on the beach in the distance. The "glass boulders" can be reached through the main entry lobby, which connects the street level to the cafeteria and surfer's kiosk, and are also accessible independently through the plaza, which serves as a main gathering space open to the public.

The museum store is located at the intermediate level of the exhibition spaces, with direct access to the entry lobby and the auditorium. The more intimate restaurant and the elevated outdoor terrace are at the top level of the museum, providing open ocean views.

At the building's southwest corner, there is a skate pool dedicated to the surfers' hangout on the plaza level and an open porch underneath, which connects to the auditorium and exhibition spaces inside the museum. This covered area provides a sheltered space for outdoor interaction, meetings and events.

The exterior of the building is textured white concrete made of aggregates from the south of France. Materials of the plaza are a progressive variation of Portuguese cobblestones paving with grass and natural vegetation. A combination of insulated glass units with clear and acid-etched layers animates the visual dynamics enhancing interior comfort. The interior of the main space is white plaster and a wooden floor provides under-floor wiring facilities. Steven Holl Architects

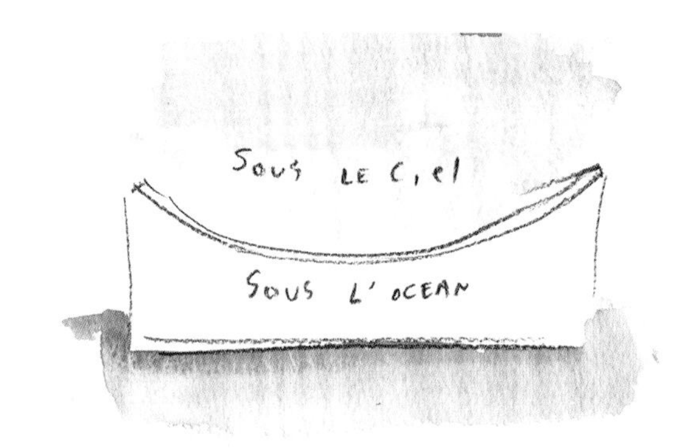

PLACE de L'OCEAN, BIARRITZ

5/23/05

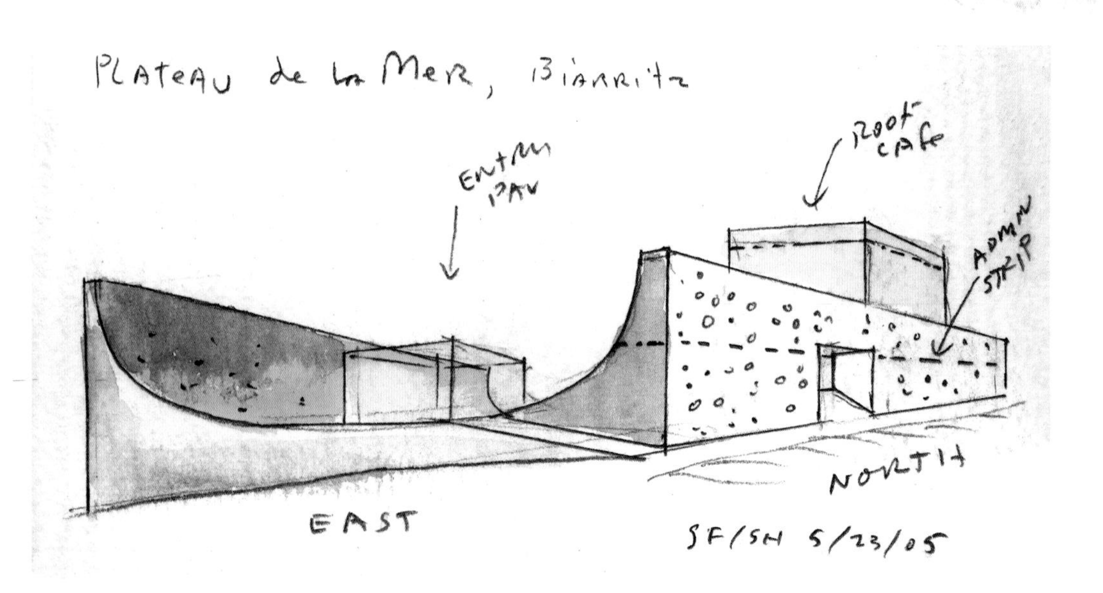

项目名称：Cité de l'Océan et du Surf (City of Ocean and Surf)
地点：Biarritz, France
建筑设计师：Solange Fabião, Steven Holl
项目建筑师：Rodolfo Dias
项目顾问：Chris McVoy
项目建筑师助理：Filipe Taboada
合作建筑师：Agence d'Architecture X.Leibar JM Seigneurin
项目团队：Francesco Bartolozzi, Christopher Brokaw, Cosimo Caggiula, Florence Guiraud, Richard Liu, Ernest Ng, Alessandro Orsini, Nelson Wilmotte, Ebbie Wisecarver, Lan Wu, Christina Yessios
DD/CD项目团队：Rüssli Architekten (Justin Rüssli, Mimi Kueh, Stephan Bieri, Björn Zepnik)
结构工程师：Betec & Vinci Construction Marseille
声效工程师：AVEL Acoustique
暖通空调工程师：Elithis
总承包商：Faura Silva, GTM Sud-Ouest Batiment
甲方：SNC Biarritz Ocean
用途：exhibition area, auditorium, restaurant, cafeteria, offices
用地面积：35,000m^2　建筑面积：4,725m^2
净楼面面积：3,800m^2　景观面积：32,500m^2
材料：concrete structure, white concrete facades, steel mullion curtain wall, stone, wood, terrazzo, steel guardrails, acoustical plaster
施工时间：2005—2011
摄影师：courtesy of architect - p.166(except as noted), p.167, p.170, p.171
©Iwan Baan(courtesy of architect) - p.162~163, p.168, p.172, p.173, p.174~175

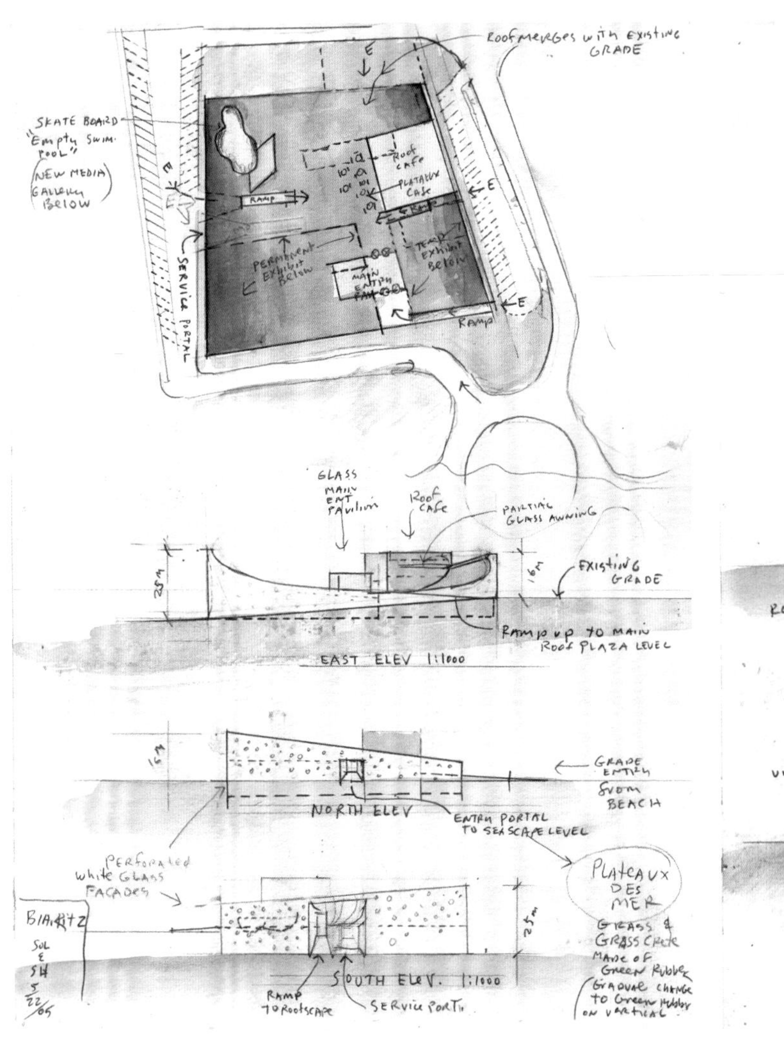

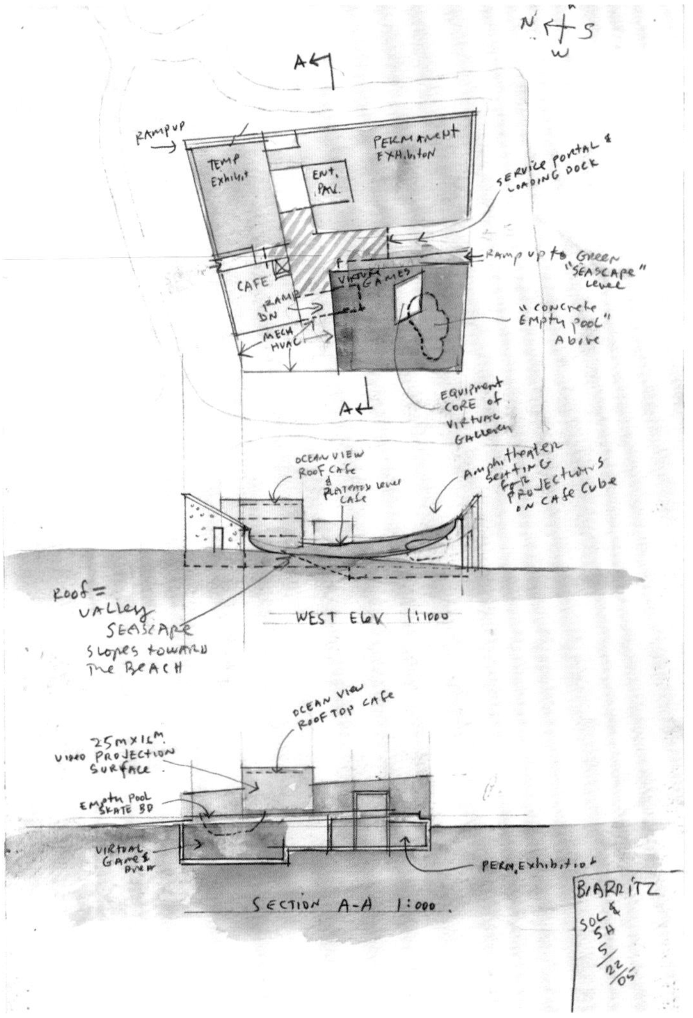

©Florent(courtesy of the architects)

©Florent(courtesy of the architects)

©Florent(courtesy of the architects)

©Florent(courtesy of the architects)

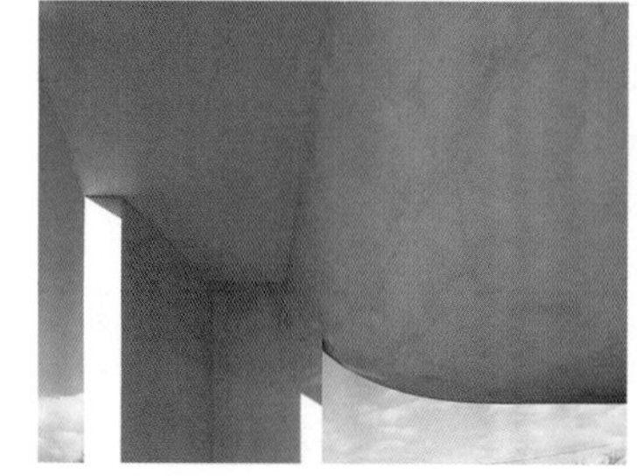

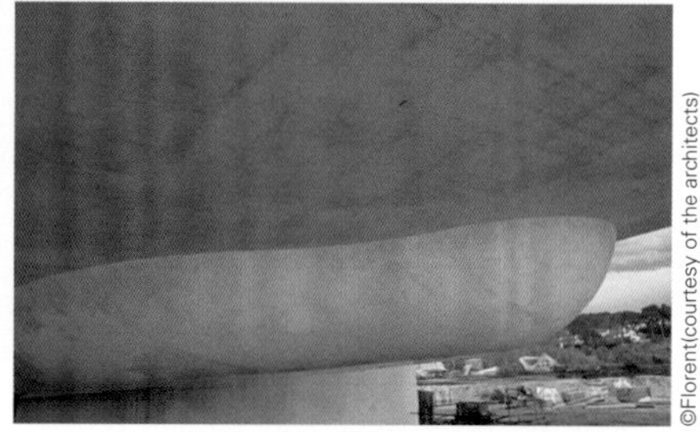

©Florent(courtesy of the architects)

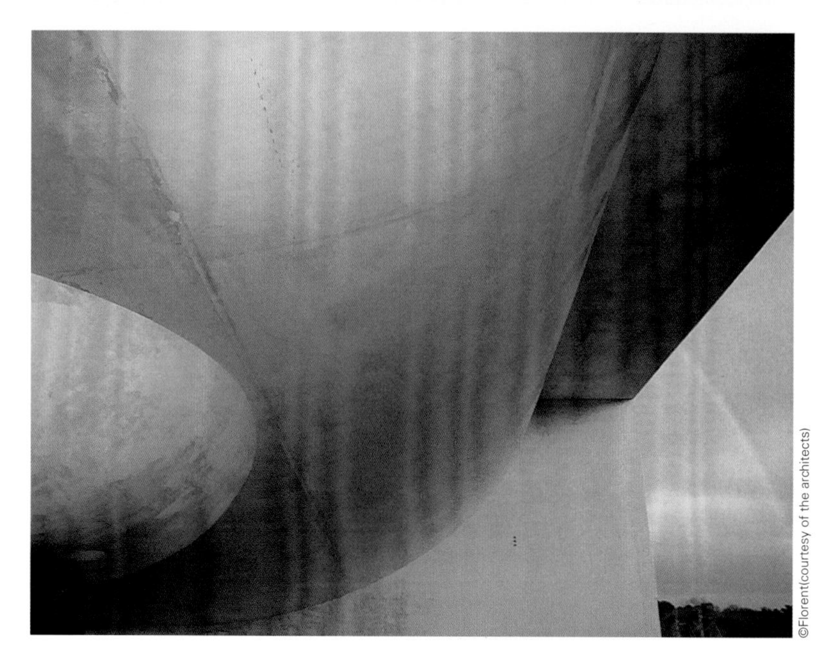

©Florent(courtesy of the architects)

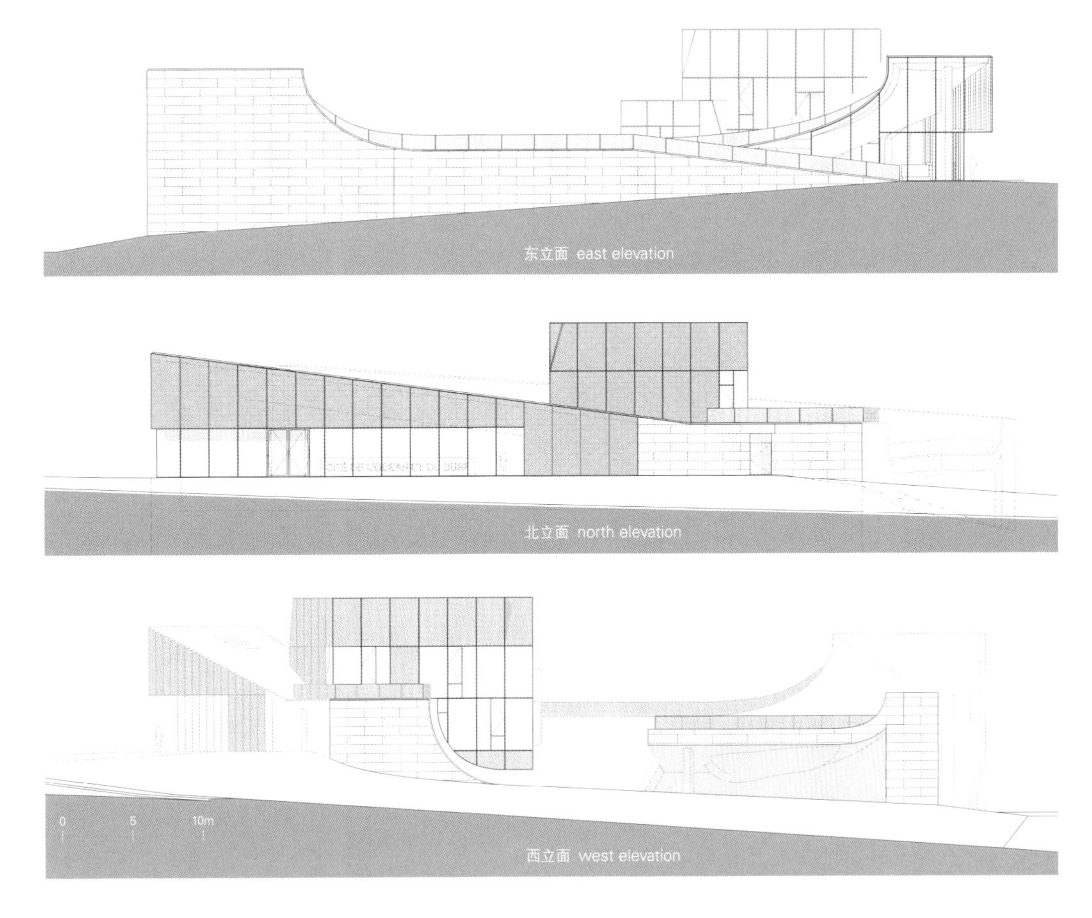
东立面 east elevation
北立面 north elevation
0
5
10m
西立面 west elevation

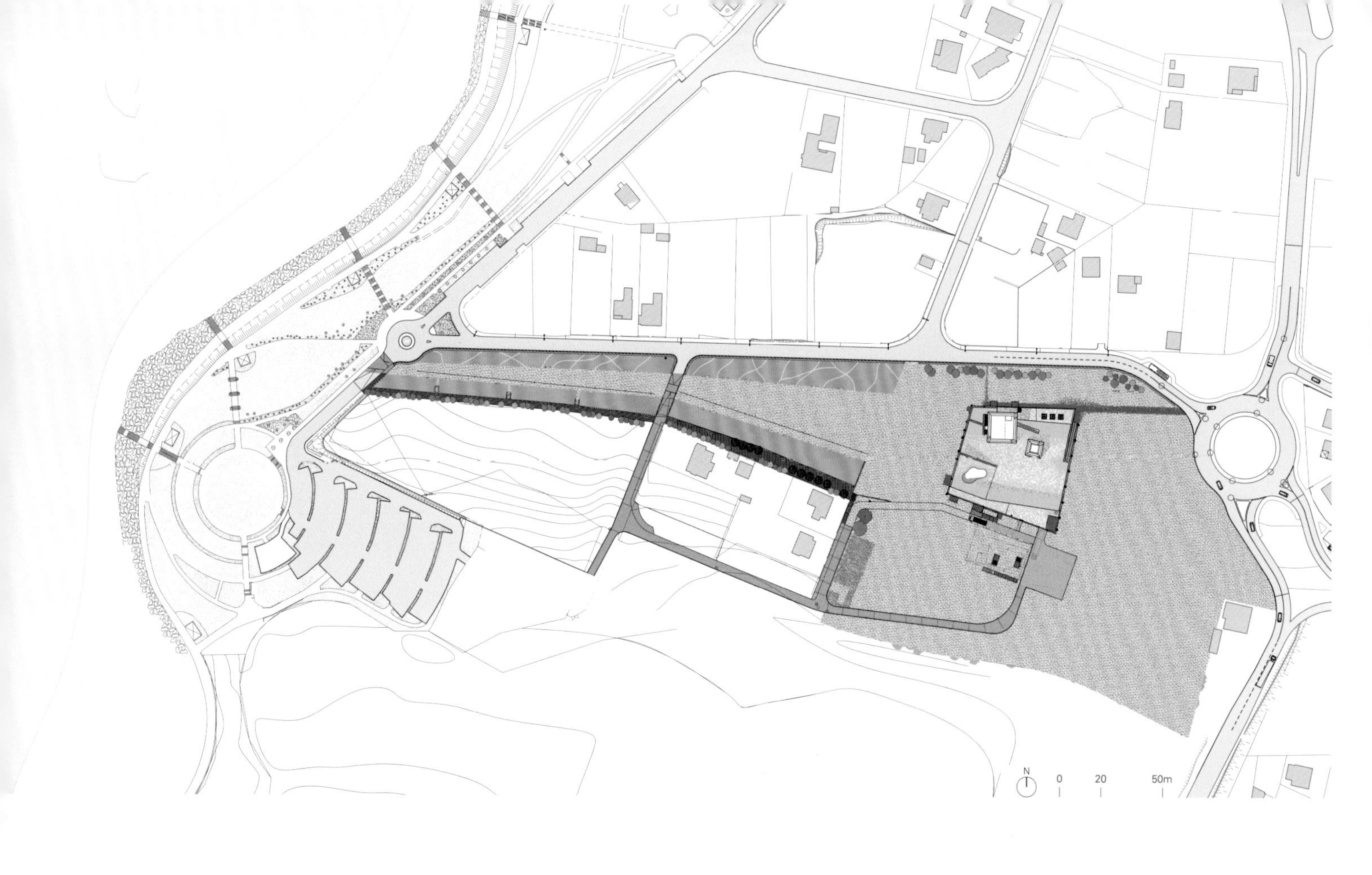
N
0
20
50m

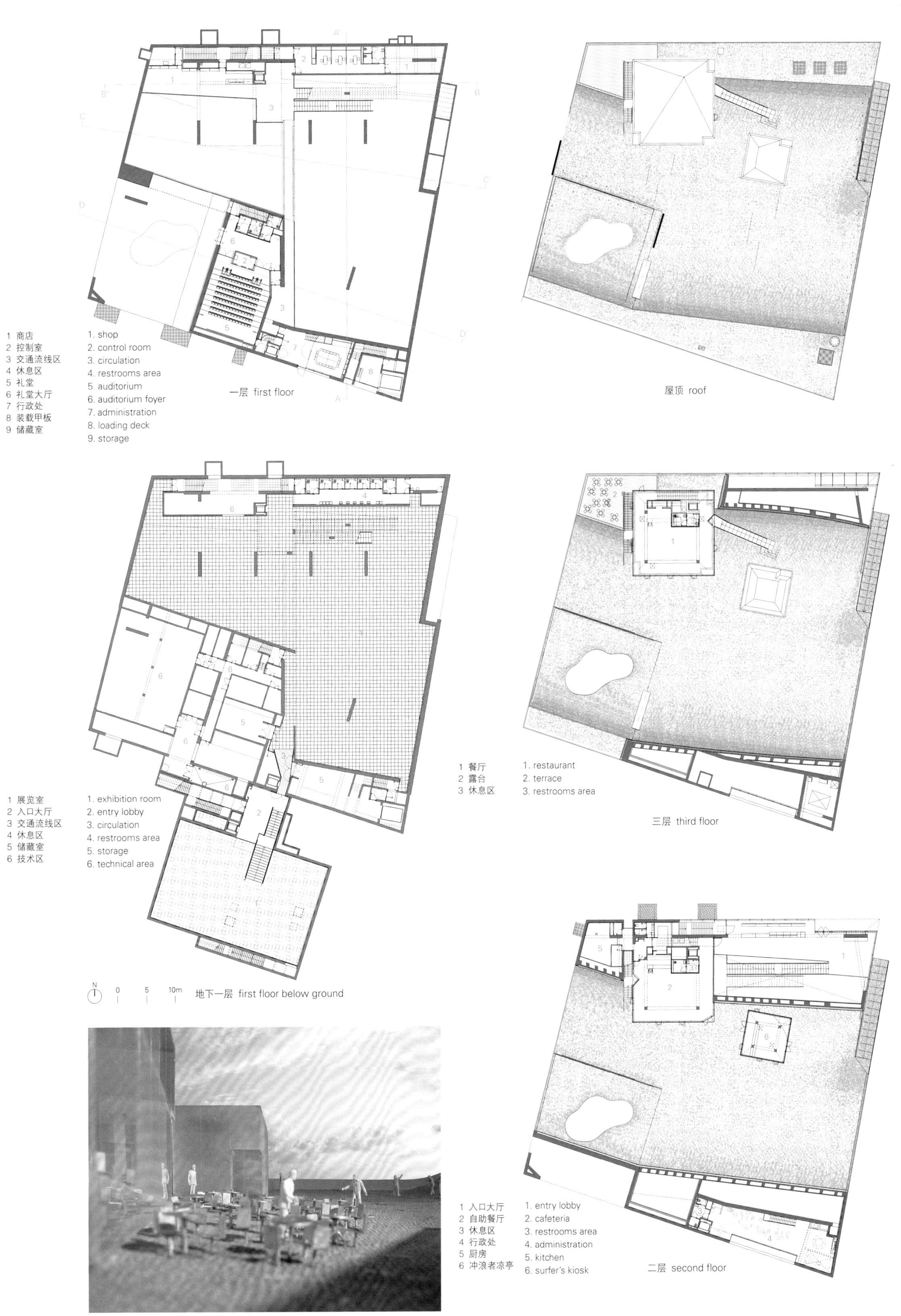
1 商店
2 控制室
3 交通流线区
4 休息区
5 礼堂
6 礼堂大厅
7 行政处
8 装载甲板
9 储藏室
1. shop
2. control room
3. circulation
4. restrooms area
5. auditorium
6. auditorium foyer
7. administration
8. loading deck
9. storage
一层 first floor
屋顶 roof
1 展览室
2 入口大厅
3 交通流线区
4 休息区
5 储藏室
6 技术区
1. exhibition room
2. entry lobby
3. circulation
4. restrooms area
5. storage
6. technical area
N
0
5
10m
地下一层 first floor below ground
1 餐厅
2 露台
3 休息区
1. restaurant
2. terrace
3. restrooms area
三层 third floor
1 入口大厅
2 自助餐厅
3 休息区
4 行政处
5 厨房
6 冲浪者凉亭
1. entry lobby
2. cafeteria
3. restrooms area
4. administration
5. kitchen
6. surfer's kiosk
二层 second floor

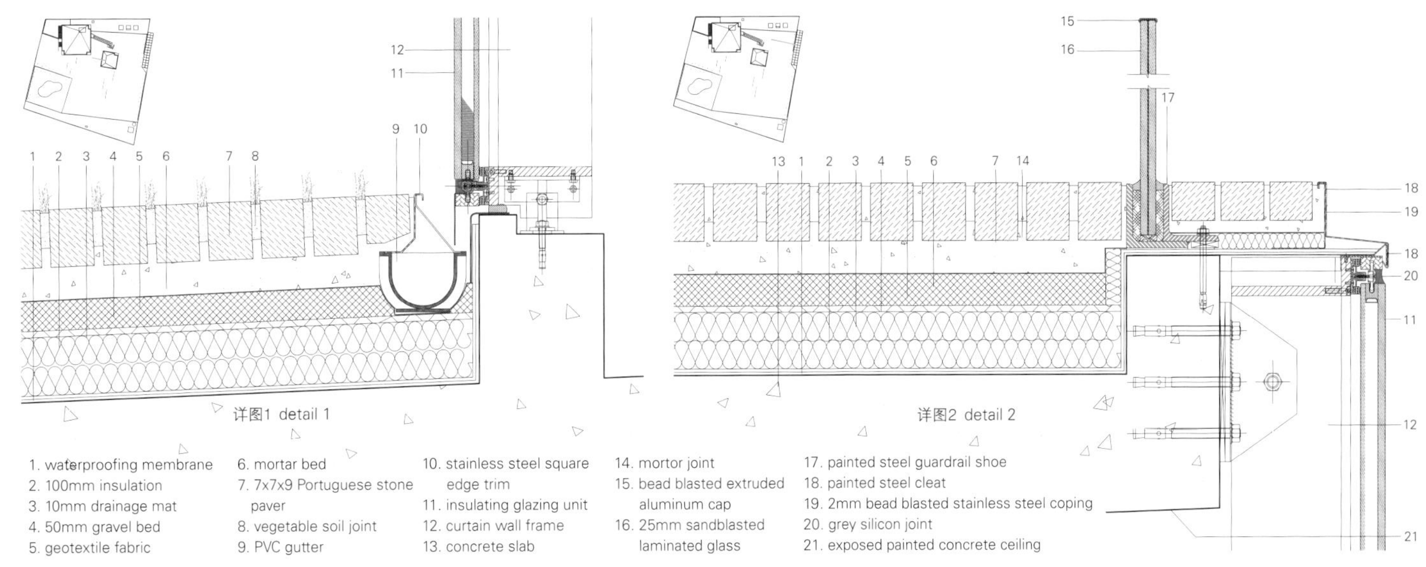

详图1 detail 1
详图2 detail 2
1. waterproofing membrane
2. 100mm insulation
3. 10mm drainage mat
4. 50mm gravel bed
5. geotextile fabric
6. mortar bed
7. 7x7x9 Portuguese stone paver
8. vegetable soil joint
9. PVC gutter
10. stainless steel square edge trim
11. insulating glazing unit
12. curtain wall frame
13. concrete slab
14. mortor joint
15. bead blasted extruded aluminum cap
16. 25mm sandblasted laminated glass
17. painted steel guardrail shoe
18. painted steel cleat
19. 2mm bead blasted stainless steel coping
20. grey silicon joint
21. exposed painted concrete ceiling

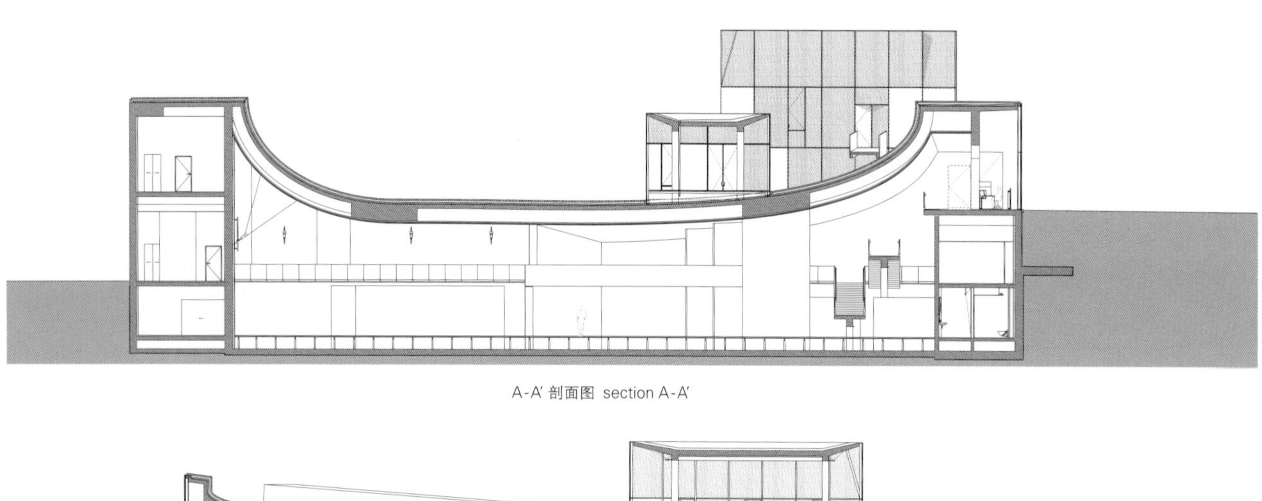

A-A′ 剖面图 section A-A′

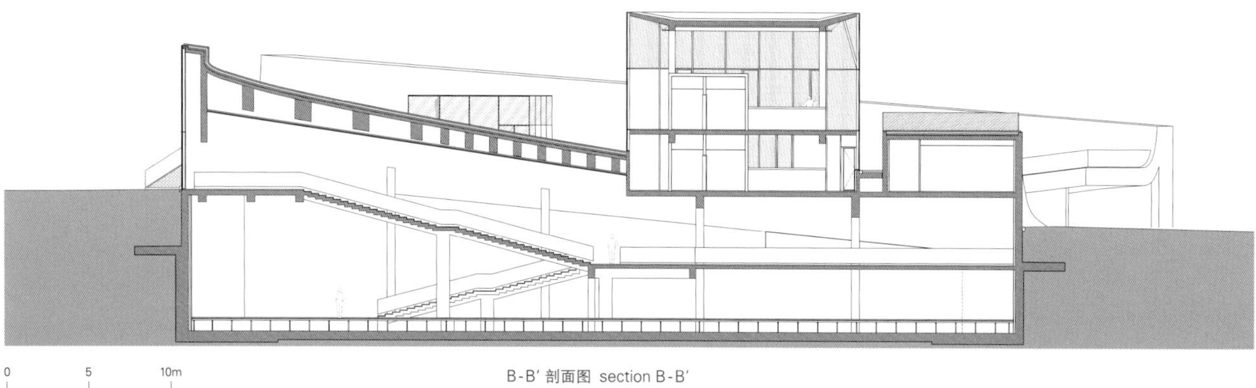

B-B′ 剖面图 section B-B′

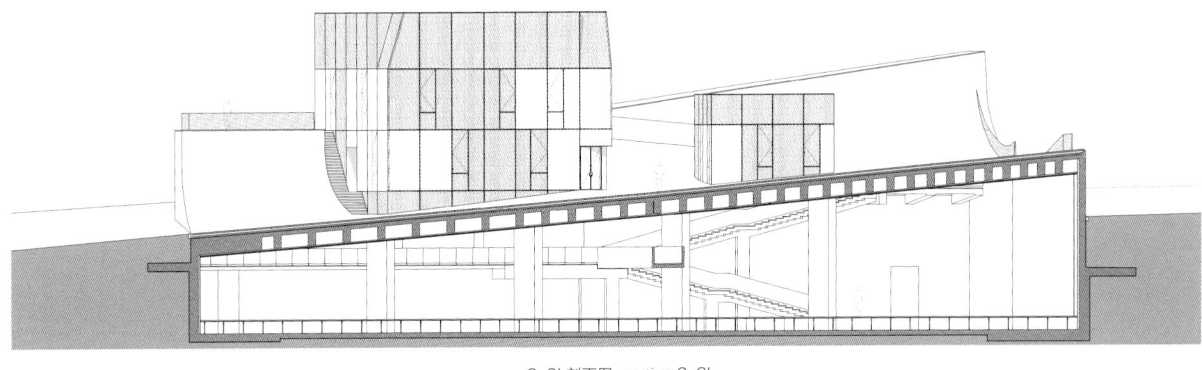

C-C′ 剖面图 section C-C′

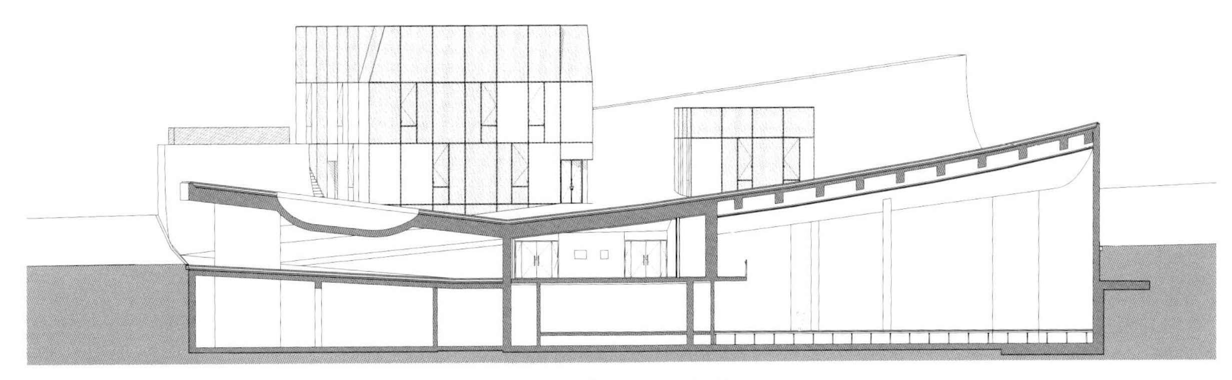

D-D′ 剖面图 section D-D′

南京四方艺术博物馆

透视法是中西绘画历史上的基本分歧。13世纪以后，西方绘画在特定的透视法中发展出了消失点，而中国画家虽然也懂得透视法，却拒绝接受单一消失点理论，继续采用"平行透视法"来描绘景观，产生一种人在画中游的效果。

新的南京四方艺术博物馆是斯蒂芬　霍尔建筑师事务所和李虎合作完成的，位于面积达120万平方米的当代国际建筑展馆（CIPEA）的入口处，坐落在中国南京苍翠优美的珍珠泉景观内。博物馆在设计上充分发掘多变的视角、空间的层次感、大片的薄雾和辽阔的水域，表现出早期国画中深深交错的空间神秘感。该博物馆由一系列"平行透视"的空间和花园墙体构成的区域组成，花园墙体的材质是竹子状的黑色混凝土，其上盘旋着一条半透明的雕塑般的画廊。一层笔直的通道到了上层逐渐变得弯曲起来。上层画廊高高地悬挑在空中，以顺时针方向展开依次排布，在正好能看到远处的南京城的位置结束。这一偏远的场地通过视觉轴线与明朝首都南京相联系而变得具有都市感。

该建筑通过其纹理、几何外形和色彩与场地紧密联系。建筑里的庭院采用从南京城中心废弃庭院回收的旧胡同砖来铺砌。博物馆的颜色是单纯的黑与白，这样就与古老的绘画和中国早期书法联系起来，同时也给内部展示的艺术品和建筑物的色彩和纹理提供了背景。之前长在场地上的直径为15cm～18cm的竹子现在被用于竹子状的混凝土花园墙上，并被渗透着色成纯黑。

一间茶楼和馆长朝南的住宅向面积为3000m²的博物馆内的展览空间致意。博物馆可以进行地热制冷和供暖，还有一个覆满景天植物的绿色屋顶以及一个由太阳能集热器构成的半透明天花板，使得30%的日光可以进入室内。此外，所有的水都被回收至庭院中的池塘里，该池塘可以将屋顶和建筑内部的水都收集起来。

Nanjing Sifang Art Museum

Perspective is the fundamental historic difference between Western and Chinese painting. After the 13th Century, Western painting developed vanishing points in fixed perspective. Chinese painters, although aware of perspective, rejected the single-vanishing point method, and instead continued producing landscapes with "parallel perspectives" in which the viewer travels within the painting.

The new Nanjing Sifang Art Museum, done in collaboration with Li Hu, is sited at the gateway to the 1,200,000m² Contemporary International Practical Exhibition of Architecture (CIPEA) master plan in the lush green landscape of the Pearl Spring near Nanjing, China. The design of the museum explores the shifting viewpoints, layers of space, and expanses of mist and water, which characterize the deep alternating spatial mysteries of early Chinese painting. The museum is formed by a field of "parallel perspective" spaces

and garden walls in black bamboo-formed concrete over which a translucent sculpture gallery hovers. The straight passages on the ground level gradually turn into the winding passage of the figure above. The upper gallery, suspended high in the air, unwraps in a clockwise turning sequence and culminates at "in-position" viewing the city of Nanjing in the distance. The meaning of this rural site becomes urban through this visual axis to the great Ming Dynasty capital city of Nanjing.

The building is anchored to its site through texture, geometrics and color. The courtyard is paved in recycled old Hutong bricks from the destroyed courtyards in the center of Nanjing. By limiting the colors of the museum to black and white, the building connects to ancient paintings and early Chinese calligraphy, but also gives a background to feature the colors and textures of the artwork and architecture to be exhibited within. Bamboo of about 15~18cm in diameter, which previously grew on the site, has been used in the bamboo-formed concrete garden walls, with a black penetrating stain.

The 3,000m² museum's flexible exhibition spaces are complimented by a Tea House and a curator's residence facing the south light. The Museum has geothermal cooling and heating, a green sedum roof, and a translucent ceiling made of solar collectors, which permit 30% of available daylight to enter the interior. In addition, all water is recycled in a pond in the courtyard that collects water from the roofs and the building. Steven Holl Architects

项目名称：Nanjing Sifang Art Museum
地点：Nanjing, China
建筑设计师：Steven Holl, Li Hu
合作者主管：Hideki Hirahara
项目建筑师：Clark Manning, Daijiro Nakayama
项目团队：Joseph Kan, JongSeo Lee, Pei Shyun Lee, Tz-Li Lin, Richard Liu, Sarah Nichols
合作建筑师：Architectural Design Institute, Nanjing University
结构顾问：Guy Nordenson and Associates
灯光设计：L'Observatoire International
甲方：Nanjing Foshou Lake Architecture and Art Developments Ltd
用途：galleries, tea room, bookstore, curator's residence
建筑面积：2,787m²
竣工时间：2011
摄影师：courtesy of the architect-p.178, p.182, p.183, p.184 bottom
©Shu He(courtesy of the architect)-p.176
©Li Hu(courtesy of the architect)-p.180 top, p.184 top, p.185 top
©Iwan Baan(courtesy of the architect)-p.180 bottom, p.185 bottom

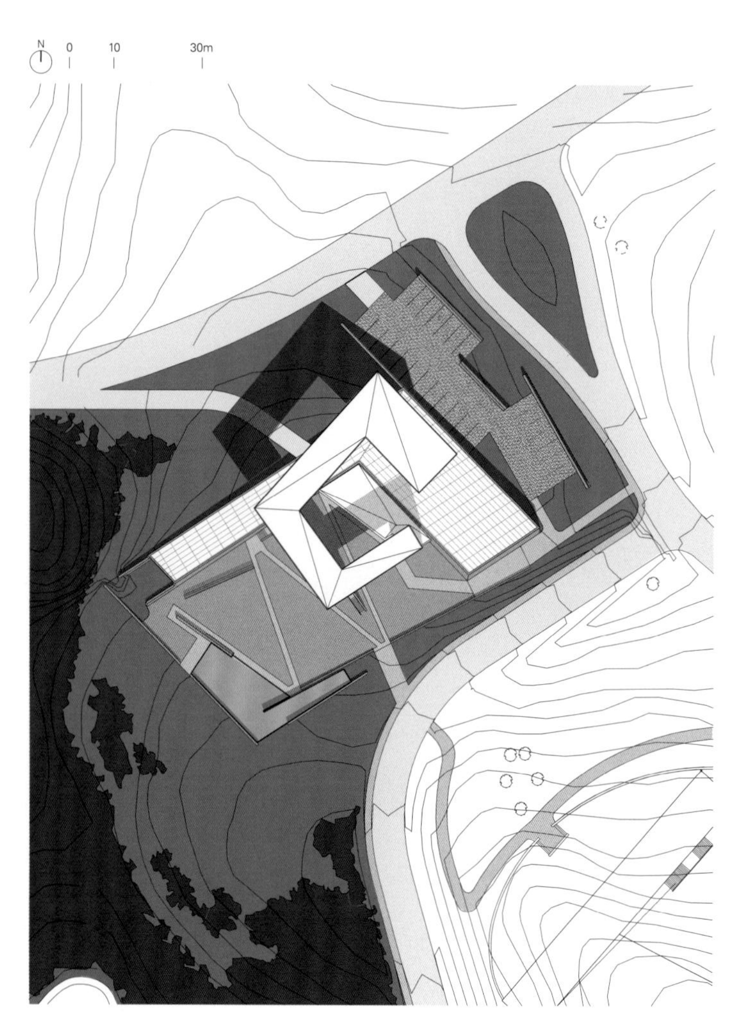

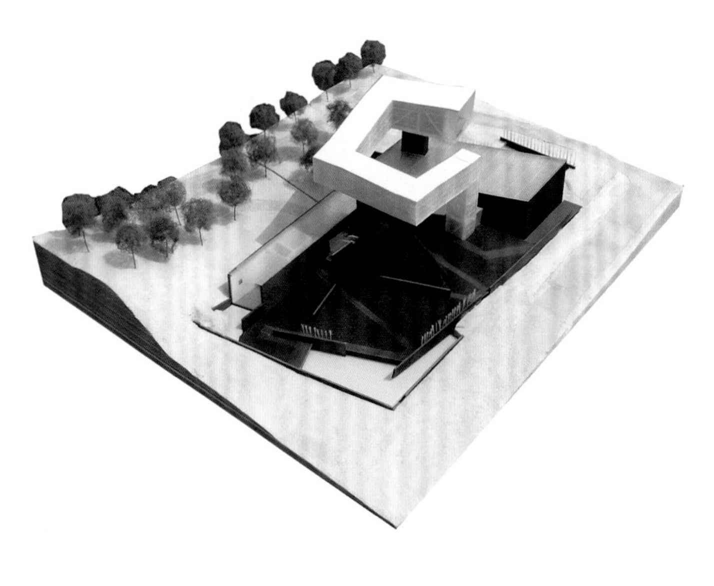

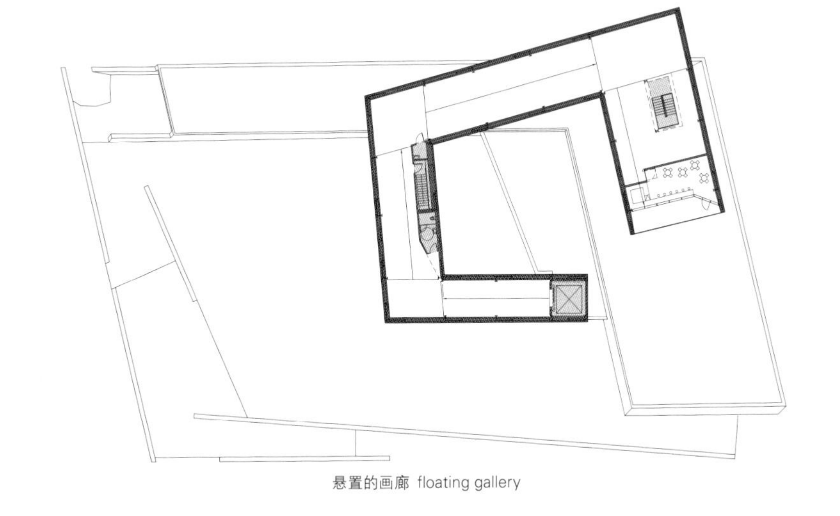

悬置的画廊 floating gallery

1 广场 2 入口 3 上方画廊 4 多功能空间 5 非正式画廊 6 接待处
7 衣帽间/储藏室 8 商店 9 咖啡厅 10 上方办公室 11 向下开放的空间 12 下方庭院 13 池塘
1. plaza 2. entrance 3. upper gallery 4. multi-purpose space 5. informal gallery 6. reception
7. coats/storage 8. shop 9. cafe 10. upper office 11. open to below 12. courtyard below 13. pond

一层 first floor

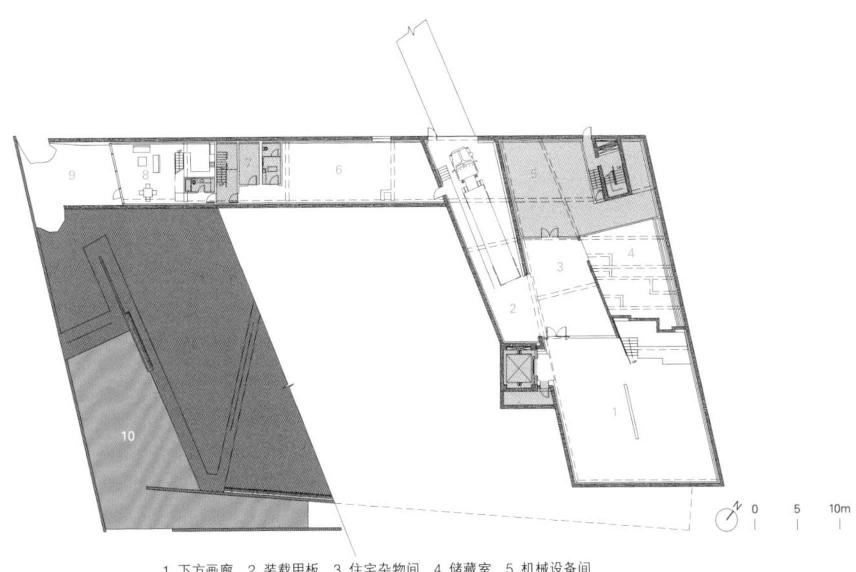

1 下方画廊 2 装载甲板 3 住宅杂物间 4 储藏室 5 机械设备间
6 行政办公室 7 杂物间门房 8 管理员住宅 9 私人庭院 10 池塘
1. lower gallery 2. loading deck 3. back of house 4. storage 5. mechanical room
6. administrative office 7. utility janitor 8. curator's residence 9. private court 10. pond

地下一层 first floor below ground

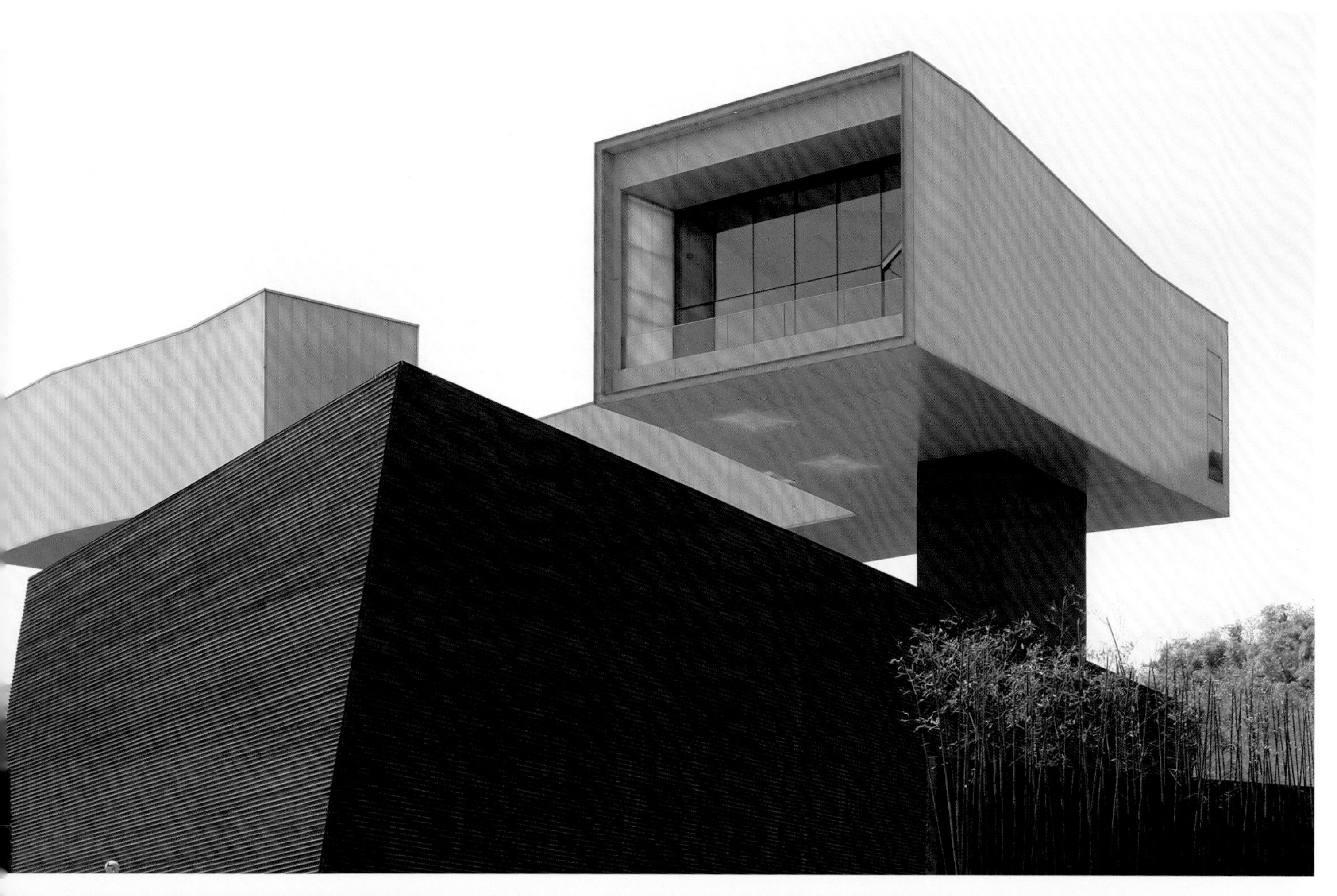

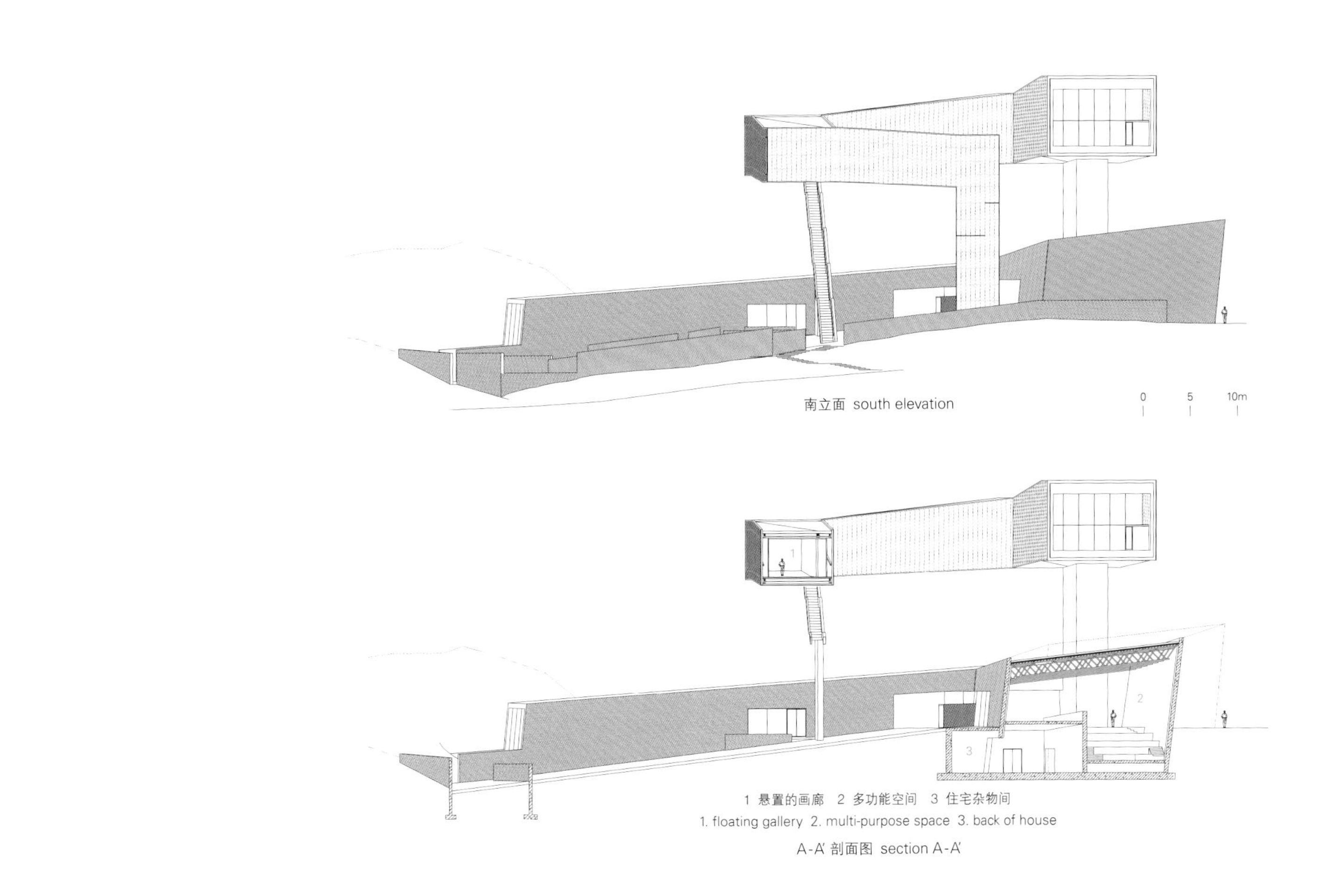

南立面 south elevation

1 悬置的画廊 2 多功能空间 3 住宅杂物间
1. floating gallery 2. multi-purpose space 3. back of house

A-A' 剖面图 section A-A'

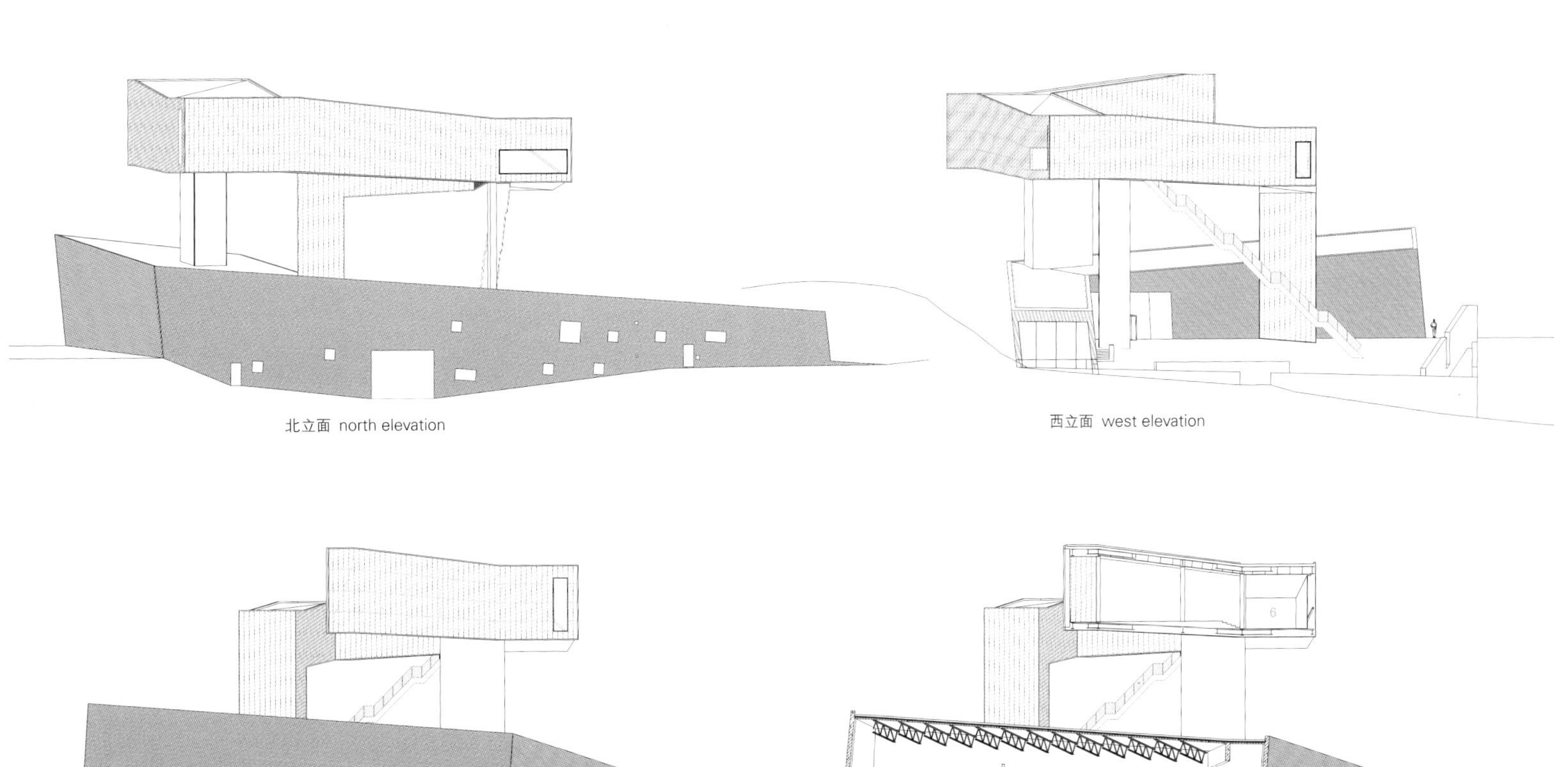

北立面 north elevation

西立面 west elevation

东立面 east elevation

1 上方画廊 2 下方画廊 3 储藏室
4 机械设备间 5 接待处 6 悬置的画廊
1. upper gallery 2. lower gallery 3. storage
4. mechanical room 5. reception 6.floating gallery

B-B' 剖面图 section B-B'

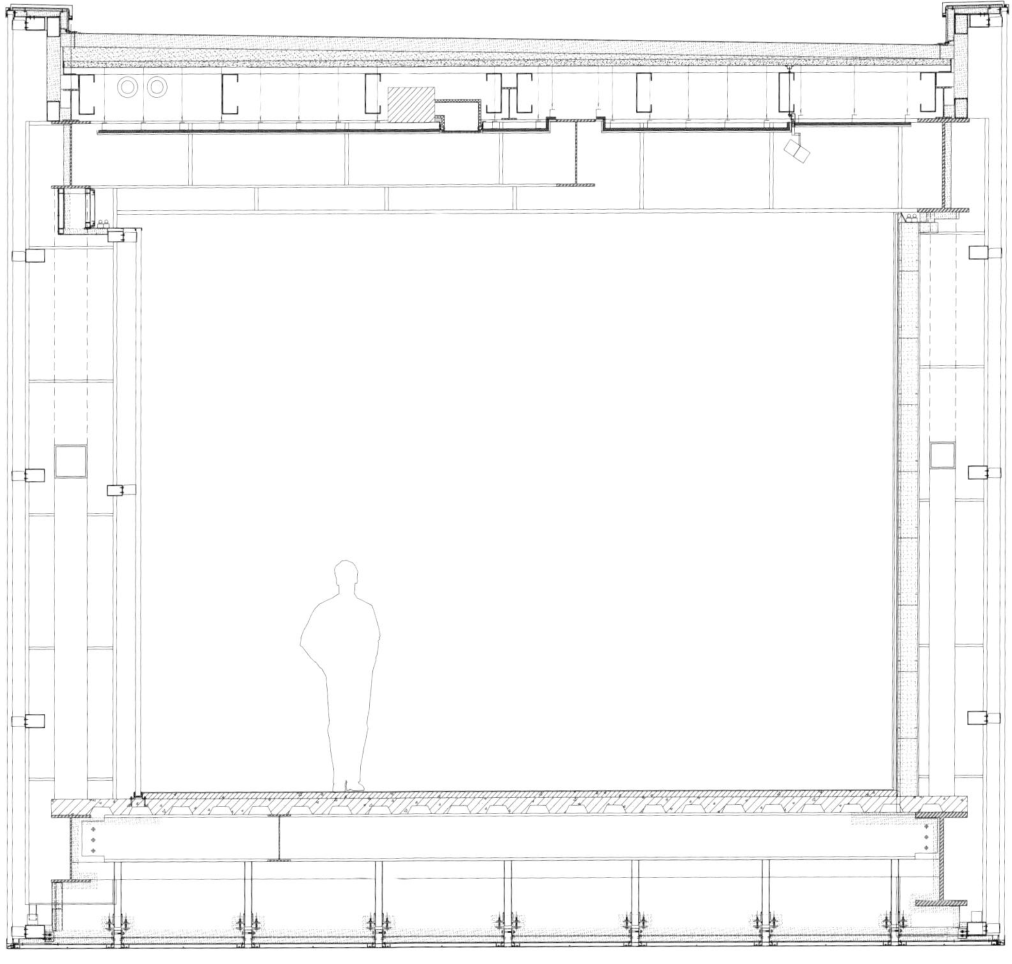

悬置的画廊剖面详图 floating gallery section detail

首尔城北画廊

这座私人画廊和住宅位于韩国首尔江北区的小山上。该项目与一家致力于音乐建筑学的研究工作室一样，都具有实验性。该建筑的基本几何形状的设计灵感来源于1967年作曲家Istvan Anhalt创作的"Modules交响乐"乐谱草稿，而这种关系是在John Cage所著的《记谱法》中发现的。

建筑有三座楼阁，一个作为入口，一个是住宅，另外一个是客房，它们看起来像是从下层连续的画廊里挤压抬升出来的。还有一片水域成为上层和下层空间的参照平面。

室内空间一直保持着"沉默"，直到它们被三座楼阁屋顶的55扇线性天窗中投射进的自然光线激活。在每一座阁楼里，五个条形玻璃窗使太阳光可以照进室内每个角落，不同季节和时段的阳光营造了不同的空间体验。天窗依照3，5，8，13，21，34，55的级数连续排列。

从楼阁向外看，景色框在倒影池中，池的周围围绕着垂直于条形天窗的花园。倒影池基座上的条形玻璃镜将耀眼的阳光反射到白色的石膏墙和下面画廊白色的花岗岩楼板上。

游客打开正门，走过几级楼梯，然后穿过竹子制成的花园墙，就到了入口庭院。在与视线水平的层面上，游客可以看到中央的池塘，并且可以在倒影中看到三座楼阁的全貌。

楼阁的内部是带有红色和黑色痕迹的竹子，天窗切入红色的竹质天花板。外部是经过特殊处理的黄铜表皮，经年累月之后将会自然地融入景观。

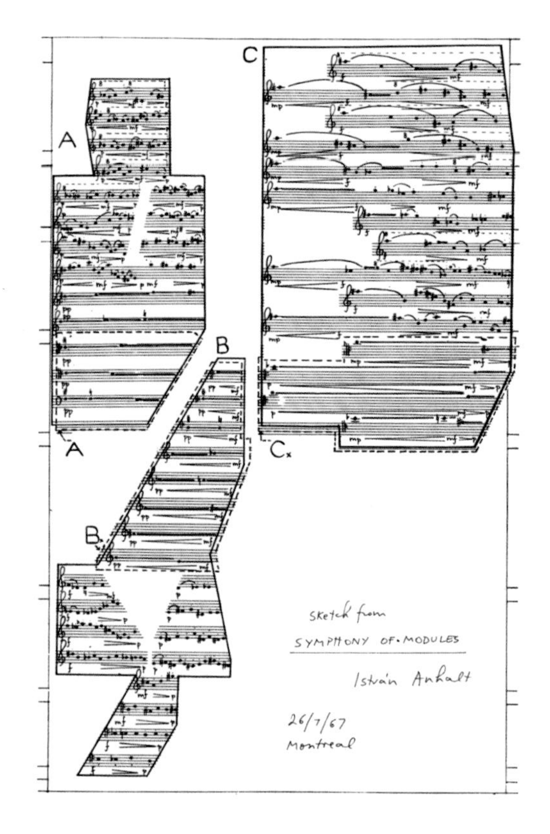

SungBuk Gallery

The private gallery and house is sited on the hills of the Kangbuk section of Seoul, Korea. The project was designed as an experiment parallel to a research studio on "the architectonics of music". The basic geometry of the building is inspired by a 1967 sketch for a music score by the composer Istvan Anhalt, "Symphony of Modules", discovered in a book by John Cage titled *Notations*.

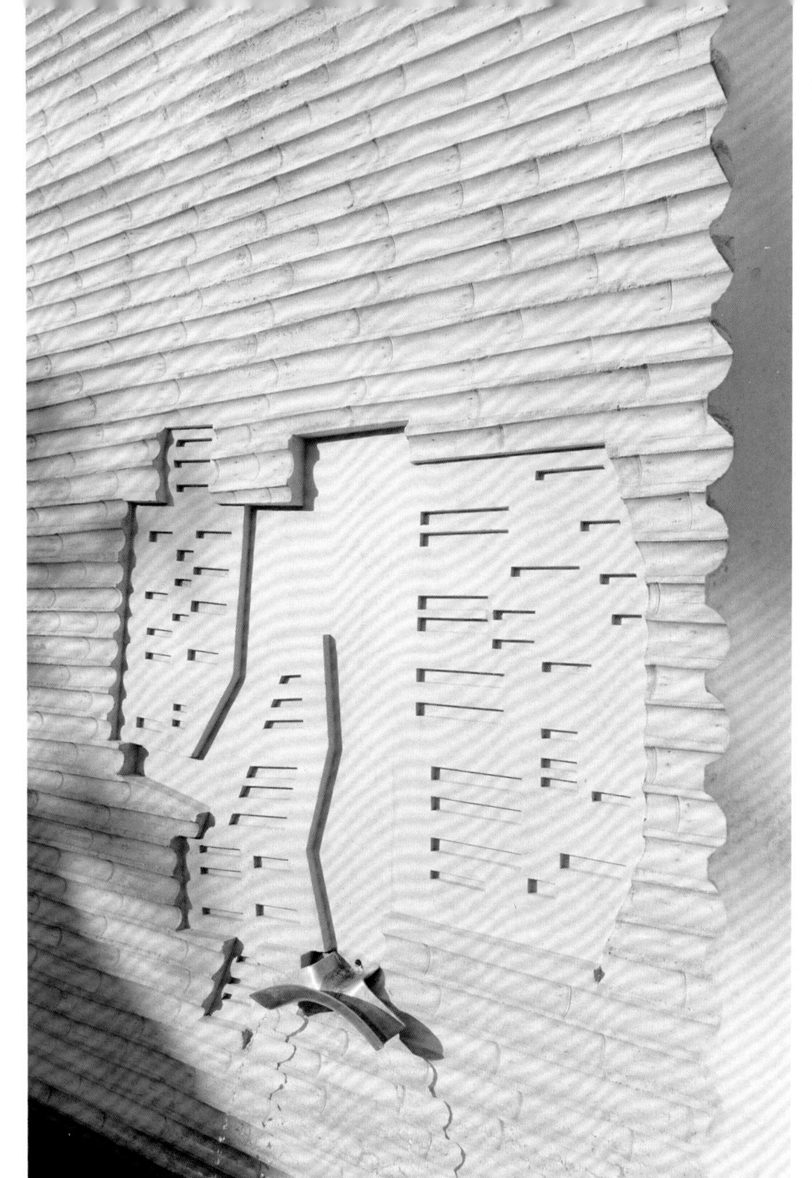

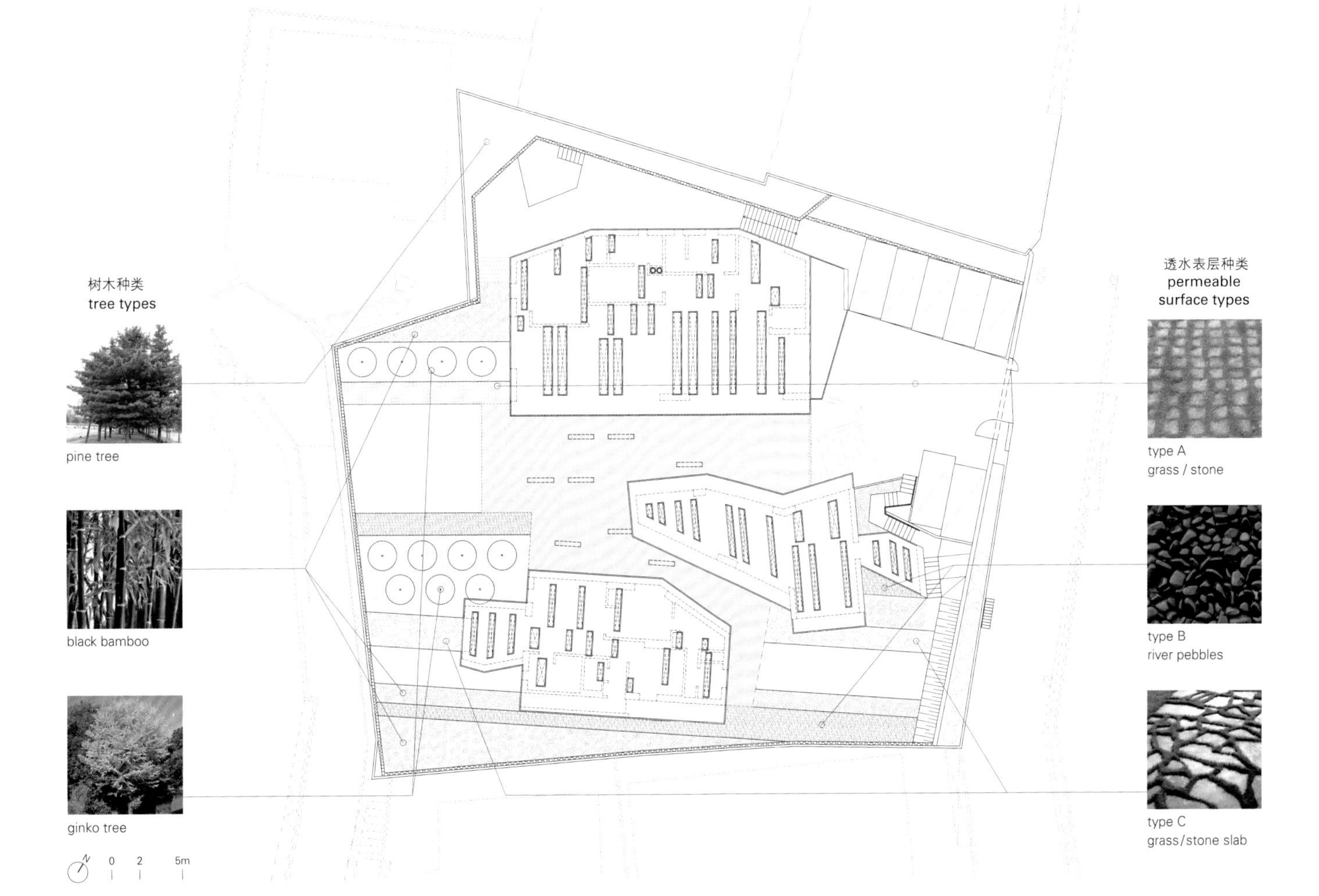
树木种类
tree types
pine tree
black bamboo
ginko tree
透水表层种类
permeable
surface types
type A
grass / stone
type B
river pebbles
type C
grass/stone slab
0
2
5m

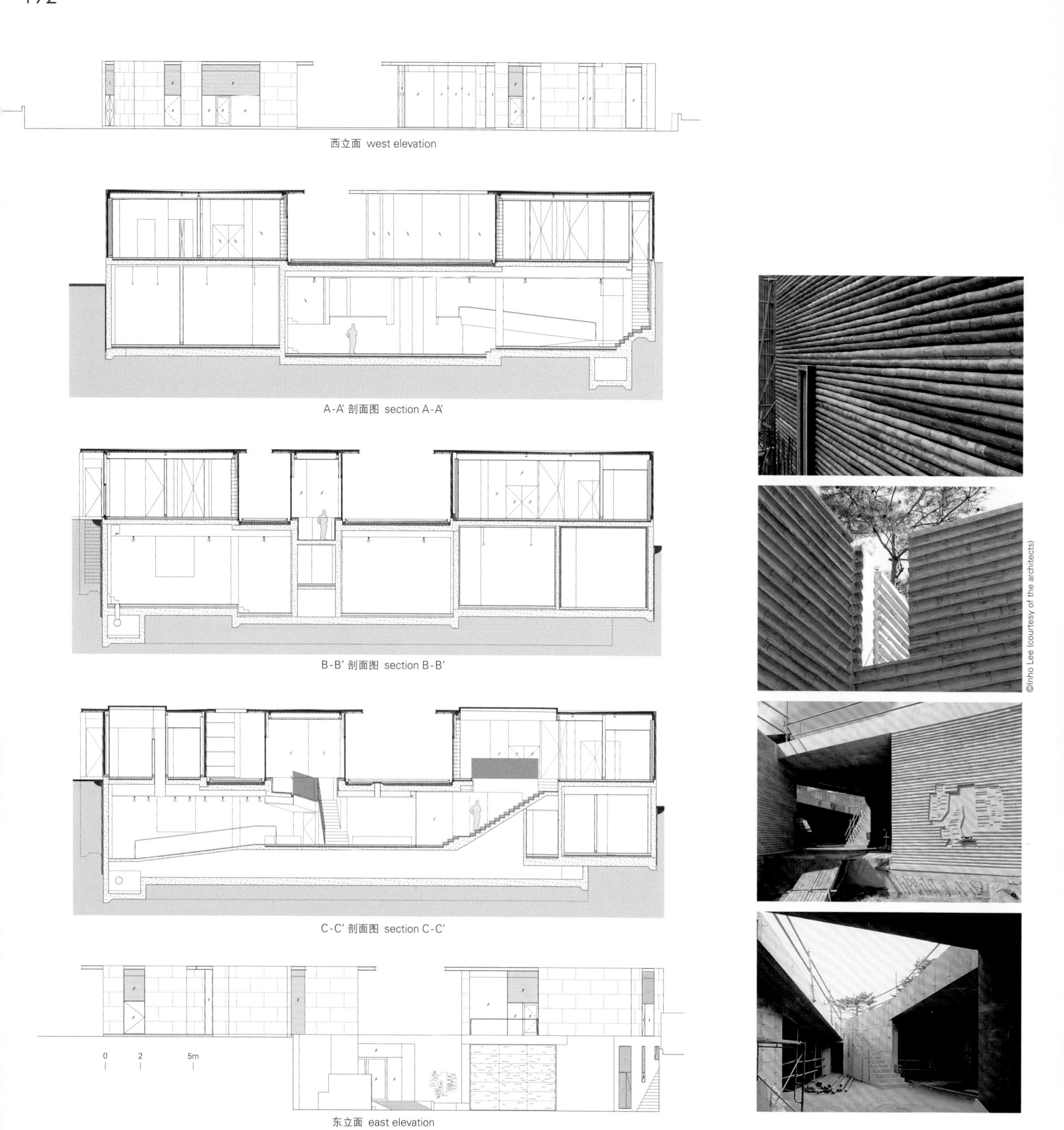

西立面 west elevation

A-A′ 剖面图 section A-A′

B-B′ 剖面图 section B-B′

C-C′ 剖面图 section C-C′

东立面 east elevation

©Inho Lee (courtesy of the architects)

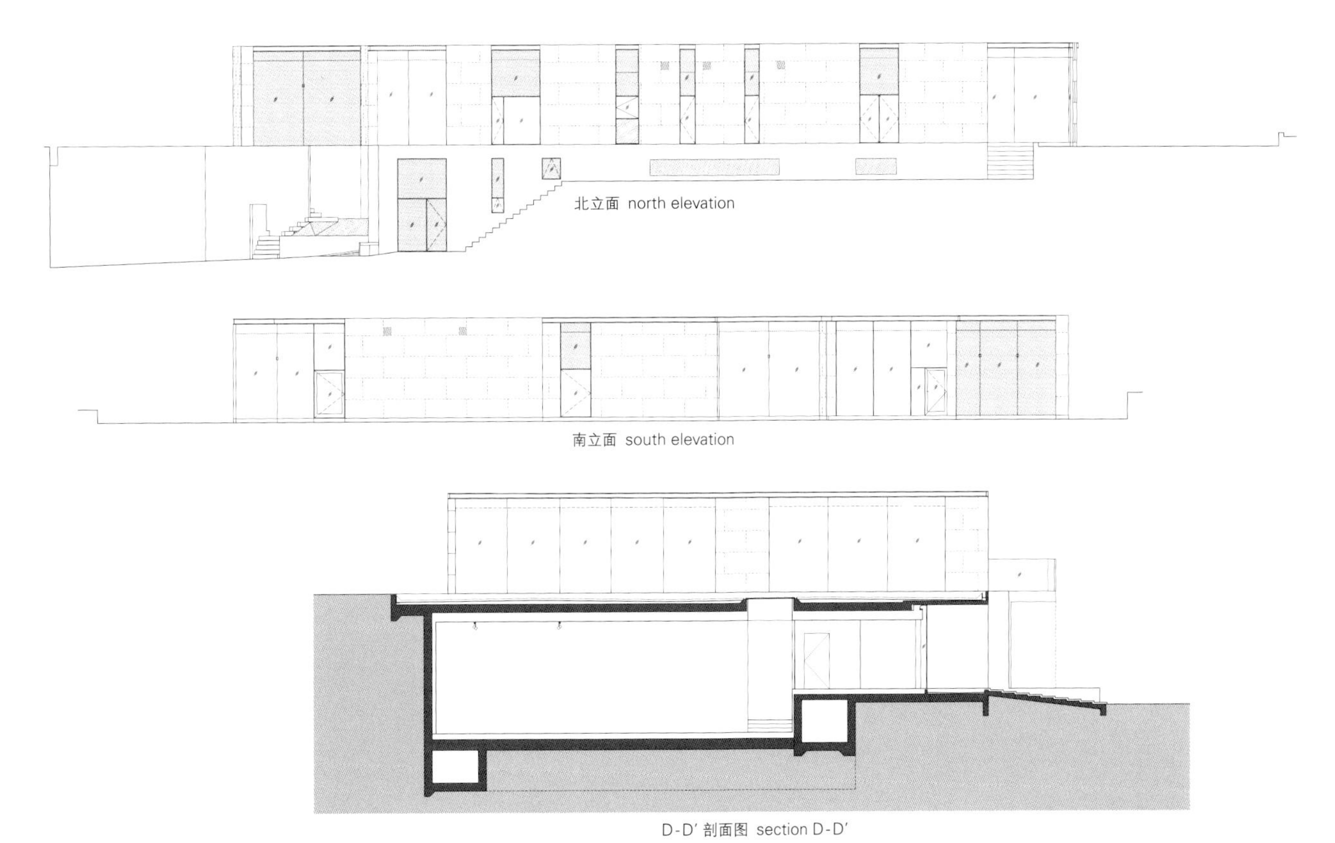

北立面 north elevation

南立面 south elevation

D-D′ 剖面图 section D-D′

项目名称：SungBuk Gallery
地点：Seoul, Korea
建筑设计师：Steven Holl
合作者主管：JongSeo Lee
项目顾问：Annette Goderbauer, Chris McVoy
项目团队：Marcus Carter, Nick Gelpi,
Fiorenza Matteoni, Rashid Satti, Dimitra Tsachrelia
当地建筑师：E.rae Architects
结构工程师：SQ Engineering
机械工程师：Buksung HVAC+R Engineering
灯光顾问：L'Observatoire International
总承包商：Jehyo
用途：residential and art gallery
总楼面面积：994m^2
结构：RC structural wall and steel
材料：exposed concrete, copper panel
设计时间：2008 竣工时间：2012
摄影师：©Iwan Baan(courtesy of the architect)
except as noted

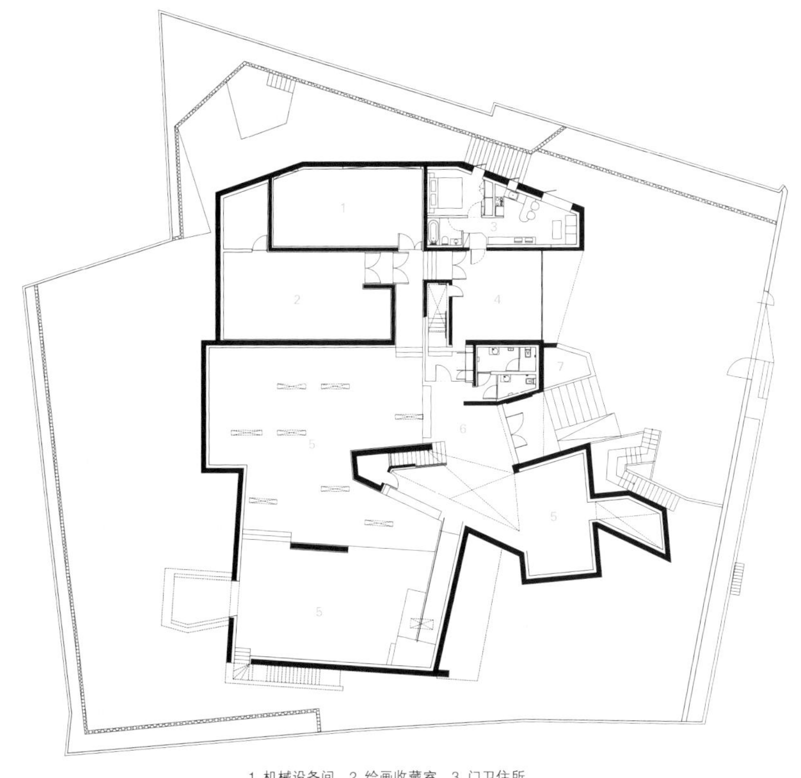

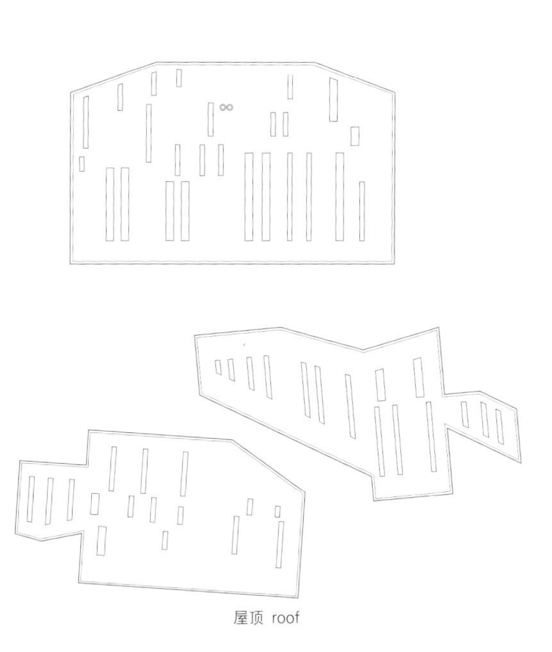

屋顶 roof

1 机械设备间 2 绘画收藏室 3 门卫住所
4 车库 5 画廊 6 门厅 7 池塘
1. mechanical room 2. painting storage 3. caretaker's house
4. garage 5. gallery 6. foyer 7. pond

一层——画廊 first floor_gallery

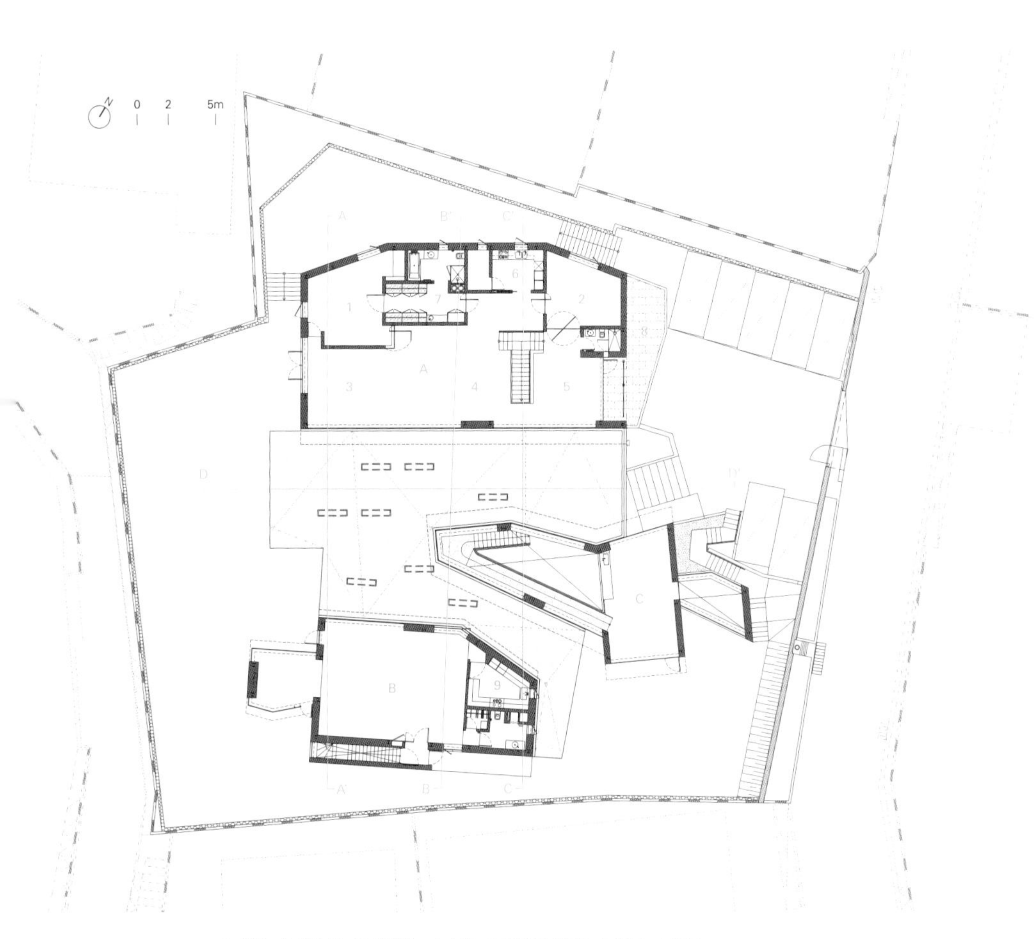

A 住所 B 活动区 C 接待处 1 主卧 2 图书馆/卧室 3 客厅 4 餐厅
5 会议室 6 厨房 7 更衣室 8 露台 9 自助厨房
A. residence B. event space C. reception 1. master bedroom 2. library / bedroom 3. living 4. dining
5. meeting room 6. kitchen 7. dressing room 8. terrace 9. service kitchen

二层——楼阁 second floor_pavilion

Three pavilions: one for entry, one for residence, and one for guest house, appear to be pushed upward from a continuous gallery level below. A sheet of water establishes the plane of reference from above and below.

The idea of inner space as silent until activated by light is realized in the cutting of 55 skylight strips in the roofs of the three pavilions. In each of the pavilions, 5 strips of clear glass allow the sunlight to turn and bend around the inner spaces, animating them according to the time of day and season. Proportions are organized around the series 3, 5, 8, 13, 21, 34, 55.

Views from within the pavilions are framed by the reflection pool which is bracketed by gardens that run perpendicular to the skylight strips. In the base of the reflecting pool, strips of glass lenses bring dappled light to the white plaster walls and white granite floor of the gallery below.

A visitor arrives through a bamboo-formed garden wall at the entry court, after opening the front door and ascending a low stair. He or she can turn to see the central pond at eye level and take in the whole of the three pavilions floating on their own reflections. The interiors of the pavilions are red and charcoal stained bamboo with the skylights cutting through a red bamboo ceiling. Exteriors are a skin of specially treated brass, which with ages will integrate naturally within the landscape. Steven Holl Architects

>>78

Rafael de La-Hoz Arquitectos

Rafael de La-Hoz was born in Spain in 1955. Graduated from the Higher Technical College of Architecture of Madrid, and went on to obtain an MDI Masters in the Polytechnic University of Madrid. Now he directs his own architectural studio and works in Spain, Portugal, Poland, Romania, Hungary, and the United Arab Emirates. Is a visiting professor at several universities, and is a member of the Editorial Council of the COAM Architectural Magazine.

>>100

ACXT Architects

Marco Suarez started his professional career in 1998. Joined ACXT-IDOM as architect and project director in 2003. The 112 Emergency building in Reus has been widely highlighted in numourous occasions including the "Green Building Challenge 2011" and the "9th Architecture Biennale of São Paulo". Developed the plan for the New Airport of Natal in Brazil.

>>14

Ai Weiwei

Ai Weiwei was born in 1957 in Beijing. Is an architect and also known as conceptual artist, photographer, curator and globally recognized human rights activist. Began his training at Beijing Film Academy in 1978 and later continued at the Parsons School of Design in New York City. His work has been exhibited around the world.

>>14

Herzog & de Meuron

Jacques Herzog[left] and Pierre de Meuron[right] studied architecture at the Swiss Federal Institute of Technology Zürich(ETH) from 1970 to 1975. Received their degree in architecture in 1975 and established Herzog & de Meuron in Basel in 1978. Have been awarded numerous prizes including "The Pritzker Architecture Prize" in 2001, the "RIBA Royal Gold Medal" and the "Premium Imperiale" in 2007. Both are visiting professors at the Harvard Graduate School of Design since 1994 (and in 1989) and professors at the ETH Zürich since 1999.

>>34

3XN

Is managed by 4 partners. Jan Ammundsen was born in 1972. Is partner and head of 3XN competition department in Copenhagen. Graduated from the Aarhus School of Architecture in 2000 and joined 3XN in 2005. Kasper Guldager Jorgensen is an innovator and developer at 3XN. Is head of GXN, an internal innovation unit established in 2007 to exploit the possibilities that arises applying latest knowledge and technology into design and architecture. Bo Boje Larsen was born in 1951. Is partner and CEO with responsibility for administration, finance, strategy and organization. Graduated from the Royal Danish Academy of Fine Arts, Copenhagen in 1977. Joined 3XN as partner in 2003. Kim Herforth Nielsen was born in 1954. Is founder, principal and artistic director of 3XN. Graduated from the Aarhus School of Architecture in 1981 and was one of the three Nielsen-founders in 1986.

>>148

Bergmeister Wolf Architekten

Gerd Bergmeister[left] was born in Brixen, Italy in 1969. After completing his architectural studies at the Leopold Franzens University in Innsbruck, established his office Gerd Bergmeister Architekten in Brixen. Michaela Wolf[left] was born in Merano, Italy in 1979. Studied architecture at the Leopold Franzens University, Architectural Association in London, and the Polytechnic University of Milan. Joined the architectural office of Gerd Bergmeister in 2006. Both began their common professional path in 2009 with the establishment of the BergmeisterWolf Architekten and were invited to give several lectures at home and abroad. Received several awards of appreciation.

>>162, 176, 188

Steven Holl Architects

Was founded in New York in 1976 and has offices in New York and Beijing. Steven Holl leads the office with partners Chris McVoy(New York) and Li Hu (Beijing). Graduated from the university of Washington and pursued architecture studies in Rome in 1970. Joined the Arhcitectural Association in London in 1976. Is recognized for his ability to blend space and light with great contextual sensitivity and to utilize the unique qualities of each project to create a concept-driven design. Specializes in seamlessly integrating new projects into contexts with particular cultural and historic importance. Is a tenured faculty member at Columbia University where he taught since 1981.

>>40

Casa Sólo Arquitectos

Was established in 1987 by Bernat Gato[left], Francesc Pernas[middle], and Roger Pernas[right]. Set their sights on building hospitals with a new design concept.

Their previous and current experience makes very strong in the field of retrofit, reuse, and renewal of existing facilities. Their expertise includes health care strategic and functional programming, as well as master plans.Their projects have always been energy efficient. Has mastered solar energy applications, and the intelligent use and management of water resources.

>>68

MGM Arquitectos

Was established by Jose Morales Sanchez^left and Juan Gonzalez Mariscal in 1987. The work has been exhibited in the Venice Biennale in 2000, 2002 and 2006. Received Swiss Architectural Award both in 2009 and 2010. Jose Morales Sanchez was born in Sevilla in 1960. Graduated from Seville School of Architecture ETSA in 1985. In 1988, received Ph.D in architecture in ETSA. Has been professor in the same university since 2004. Sara de Giles Bubois^right was born in Seville in 1972. Grduated from ETSA in 1998 and joined MGM Arquitectos. Has been associate lecturer of architectural projects department ETSA in Seville since 2006.

>>124

X-TU Architects

Was established by Anouk Legendre^left and Nicolas Desmazieres^right. Both graduated from school of Architecture paris-La Vilette and became officially certificated architects in 1988. Has built up a refined and precise language for sincere placeness, serene and light volumes, and therefore withholding a unique force of impact.

>>142

Stein Halvorsen AS Sivilarkitekter MNAL

Was established in 1996. Received a number of awards in architectural competitions. Many of the projects are public spaces including cultural facilities, art galleries, churches, court houses and fire stations. Received the Norwegian Building prize in 2001 and North Norway's architectural prize in 2002.

>>136

Estudio Lamazeta

Was co-founded by Maria Moreno Garcia, Lucia Manresa Casado and David Mata Sanchez in 2009. Conceive the office as an open space where the interdisciplinary transfer (from art to architecture, from sociology to city, etc.) allows us to face more extensive and complete way of the current scene of architecture and design. Has worked in various kinds of projects, competitions, collaborations as well as teaching and research. In 2011, worked in collaboration with the investigation project 'Red Nacional de Silos. Integration into the urban reality of Andalusia and its reuse for new typologies. Junta de Andalucía'.

>>28, 48

C.F. Møller Architects

Is one of Scandinavia's oldest and largest architectural practices founded in 1924 by C.F Møller. Is a partnership owned and managed by 9 partners; Julian Weyer, Mads Mandrup, Tom Danielsen, Anna Maria Indrio, Klavs Hyttel, Lars Kirkegaard, Lone Wiggers, Mads Moller and Klaus Toustrup. Over the years, has won a large number of national and international competitions. Also has been exhibited locally as well as internationally, at RIBA in London, the Venice Biennale, and the Danish Cultural Institute in Beijing.

>>58

Pinearq

Is a studio dedicated to qualified technical services in architecture and urbanism with a complete team since 1997. Albert de Pineda Alvarez, the founder of Pinearq graduated from ETSAB in 1980. Specialized in health care architecture, designing hospitals and nursing homes, both public and private in Spain. Has given many conferences at several universities and congress. Is currently teaching at the Open University of Catalunya.

Human Wu

Works in Basel, Switzerland as an architect. After graduating from South China University of Technology and Harvard Graduate School of Design, he worked in New York until he moves to Switzerland. Has written for magazines including Time+Architecture (Shanghai) and MONU (Rotterdam), and also in his own blog Human's Scribbles.

Andrew Tang

Received his architecture degree at Institute of Design(IIT), Chicago in 1996. In the same year, he won the Jerrold Loebl Prize. Has worked around the world with many prominent figures and reputable offices such as West 8, MVRDV, Maxwan, and Architecten Cie before starting his own design practice, Tanglobe. Has also written many articles and lectured in many institutions and conferences with a strong passion for architecture.

>>132

Paredes Pedrosa Arquitectos

Was estabilshed by Ángela García de Paredes[top], and Ignacio Pedrosa[bottom] in 1990. Both graduated from Madrid Architecture School where they work as the architectural design professors. Dedicated to competitions, and project and urban design mainly for public buidings related to cultural equipment. Their works have been awarded with the Spanish Architecture award 2007, Madrid Architecture award, and Madrid Institute of Architects Award. Participate in exhibitions in FAD Barcelona, Spanish Architecture Biennales, Venice Biennale, RIBA London, Sao Paulo Biennale.

Simone Corda

Is an architect and PhD candidate based in Sydney, Australia. Explores the themes of contemporary architecture through researches and projects at different scales and cross sectors. Referring to the architecture of Australia and New Zealand, he is currently focusing on the flexibility of housing as the key concept for sustainability. Part of his PhD thesis about Glenn Murcutt's work has been already published in the Italian magazine Area. Contributes to the Faculty of Architecture at the University of Cagliari enthusiatically with regular seminars and lectures at the Faculty of Architecture in Alghero, C.N.R. National Center of Research and Festarch event.

Alison Killing

Is an architect and urban designer based in Rotterdam, the Netherlands. Has written for several architecture and design magazines in the UK, contributing features and reviews to Blueprint and Icon and editing the research section of Middle East Art Design and Architecture. Most recently, she has worked as a correspondent for the online sustainability magazine Worldchanging. Has an eclectic design background, ranging from complex geometry and structural engineering, to humanitarian practice, to architecture and urban design and has worked internationally in the UK and the Netherlands, but also more widely in Europe, Switzerland, China and Russia.

图书在版编目(CIP)数据

新医疗建筑：汉英对照/韩国C3出版公社编；孙佳等译. —大连：大连理工大学出版社，2012.10
（C3建筑立场系列丛书；20）
ISBN 978-7-5611-7328-2

Ⅰ. ①新… Ⅱ. ①韩… ②孙… Ⅲ. ①医院－建筑设计－汉、英 Ⅳ. ①TU246.1

中国版本图书馆CIP数据核字（2012）第225418号

出版发行：大连理工大学出版社
（地址：大连市软件园路80号　邮编：116023）
印　　刷：精一印刷（深圳）有限公司
幅面尺寸：225mm×300mm
印　　张：12.75
出版时间：2012年10月第1版
印刷时间：2012年10月第1次印刷
出 版 人：金英伟
统　　筹：房　磊
责任编辑：张昕焱
封面设计：王志峰
责任校对：崔玉玲

书　　号：ISBN 978-7-5611-7328-2
定　　价：228.00元

发　行：0411-84708842
传　真：0411-84701466
E-mail: 12282980@qq.com
URL: http://www.dutp.cn